REFRIGERATION TABLES WITH CHARTS

(S.I. Units)

J.K. GUPTA

R.S. KHURMI

S. CHAND PUBLISHING

S Chand And Company Limited

(ISO 9001 Certified Company)

S Chand And Company Limited

(ISO 9001 Certified Company)

Head Office: D-92, Sector–2, Noida – 201301, U.P. (India), Ph. 91-120-4682700

Registered Office: A-27, 2nd Floor, Mohan Co-operative Industrial Estate, New Delhi – 110 044, Phone: 011-49731800

www.schandpublishing.com; e-mail: **info@schandpublishing.com**

Marketing Offices:

Chennai	:	Ph: 23632120; chennai@schandpublishing.com
Guwahati	:	Ph: 2738811, 2735640; guwahati@schandpublishing.com
Hyderabad	:	Ph: 40186018; hyderabad@schandpublishing.com
Jalandhar	:	Ph: 4645630; jalandhar@schandpublishing.com
Kolkata	:	Ph: 23357458, 23353914; kolkata@schandpublishing.com
Lucknow	:	Ph: 4003633; lucknow@schandpublishing.com
Mumbai	:	Ph: 25000297; mumbai@schandpublishing.com
Patna	:	Ph: 4011400; patna@schandpublishing.com

First Edition 2007

Subsequent Editions and Reprints 2008, 2009, 2010

Third Revised Edition 2011

Reprints 2012, 2013, 2014, 2015, 2016, 2018 (Twice), 2019 (Thrice), 2022, 2023, 2024

Reprint 2026

ISBN: 978-81-219-2829-8 **Product Code:** H3RTC62ACRF10ENAC11R

PRINTED IN INDIA

By Shri Shyamlal Printing Press Private Limited, Plot 20/4, Site-IV, Industrial Area Sahibabad, Ghaziabad – 201 010 and Published by S Chand And Company Limited, A-27, 2nd Floor, Mohan Co-operative Industrial Estate, New Delhi – 110 044.

PREFACE TO THE THIRD EDITION

We feel satisfied in presenting the revised edition of this small booklet. The mistakes, which had crept in the previous edition, have been eliminated. The present edition has been thoroughly revised and brought up-to-date by adding six new tables and three new charts, to make more useful for the students. We are highly thankful to Er. Sanjiv Kumar Gupta, B.Tech. (Mechanical Engineering), M.S. (Manufacturing Management) for his help in preparing the revised edition.

Although every care been taken to check mistakes and misprints in the colossal figure work, yet it is difficult to claim perfection. Any errors, omissions and suggestions for the improvement of this booklet, brought to our notice, will be thankfully acknowledged and incorporated in the next edition.

J.K. GUPTA
R.S. KHURMI

CONTENTS

Tables

Charts

TABLE 1

Dry Saturated Steam (Pressure based)

Pressure in bar (p)	Saturation temperature in°C (t)	Specific volume of steam in m^3/kg (v_g)	Specific enthalpy in kJ/kg			Specific entropy in kJ/kg K		
			Water (h_f)	Evaporation (h_{fg})	Steam (h_g)	Water (s_f)	Evaporation (s_{fg})	Steam (s_g)
0.00611	**0.000**	**206.16**	**0.8**	**2501.6**	**2501.6**	**0.000**	**9.158**	**9.158**
0.02	17.51	67.012	73.5	2460.2	2533.7	0.261	8.464	8.725
0.04	28.98	34.803	121.4	2433.1	2554.5	0.423	8.053	8.476
0.06	36.18	23.741	151.5	2416.0	2567.5	0.521	7.810	8.331
0.08	41.54	18.104	173.9	2403.2	2577.1	0.593	7.637	8.230
0.10	45.83	14.674	191.8	2392.9	2584.8	0.649	7.502	8.151
0.12	49.45	12.361	206.9	2384.2	2591.2	0.696	7.391	8.087
0.14	52.58	10.693	220.0	2376.7	2596.7	0.737	7.297	8.033
0.16	55.34	9.4324	231.6	2370.0	2601.6	0.772	7.215	7.987
0.18	57.83	8.4446	242.0	2363.9	2605.9	0.804	7.142	7.946
0.20	60.09	7.6492	251.5	2358.4	2609.9	0.832	7.077	7.909
0.25	64.99	6.2040	272.0	2346.4	2618.4	0.893	6.939	7.832
0.30	69.13	5.229	289.3	2336.1	2625.4	0.944	6.825	7.769
0.35	72.71	4.5255	304.3	2327.2	2631.5	0.988	6.729	7.717
0.40	75.89	3.9932	317.7	2319.2	2636.9	1.026	6.645	7.671
0.45	78.74	3.5761	329.6	2312.0	2641.6	1.060	6.570	7.630
0.50	81.35	3.2401	340.6	2305.4	2646.0	1.091	6.504	7.595
0.60	85.95	2.7317	359.9	2293.6	2653.5	1.146	6.387	7.533
0.70	89.96	2.3647	376.8	2283.3	2660.1	1.192	6.288	7.480
0.80	93.51	2.0869	391.7	2274.0	2665.7	1.233	6.202	7.435
0.90	96.71	1.8691	405.2	2265.6	2670.9	1.270	6.126	7.395
1.00	99.63	1.6937	417.5	2257.9	2675.4	1.303	6.057	7.360
1.013	**100.00**	**1.6730**	**419.1**	**2256.9**	**2676.0**	**1.307**	**6.048**	**7.355**
1.20	104.8	1.4281	439.4	2244.1	2683.4	1.361	5.937	7.298
1.40	109.3	1.2363	458.4	2231.9	2690.3	1.411	5.836	7.247
1.60	113.3	1.0911	475.4	2220.8	2696.2	1.455	5.747	7.202
1.80	116.9	0.9772	490.7	2210.8	2701.5	1.494	5.668	7.162
2.00	120.2	0.8854	504.7	2201.6	2706.3	1.530	5.597	7.127
2.50	127.4	0.7184	535.4	2181.0	2716.4	1.607	5.445	7.052
3.00	133.5	0.6055	561.4	2163.2	2724.7	1.672	5.319	6.991
3.50	138.9	0.5240	584.3	2147.3	2731.6	1.727	5.212	6.939
4.0	143.6	0.4622	604.7	2132.9	2737.6	1.776	5.118	6.894
4.5	147.9	0.4137	623.2	2119.7	2742.9	1.820	5.034	6.855
5.0	151.9	0.3747	640.1	2107.4	2747.5	1.860	4.959	6.819

p	t	v_g	h_f	h_{fg}	h_g	s_f	s_{fg}	s_g
6.0	158.8	0.3155	670.4	2085.1	2755.5	1.931	4.827	6.758
7.0	165.0	0.2727	697.1	2064.9	2762.0	1.992	4.713	6.705
8.0	170.4	0.2403	720.9	2046.6	2767.5	2.046	4.614	6.660
9.0	175.4	0.2148	742.6	2029.5	2772.1	2.094	4.525	6.619
10.0	179.9	0.1943	762.6	2013.6	2776.2	2.138	4.445	6.583
12.0	188.0	0.1632	798.4	1984.3	2782.7	2.216	4.303	6.519
14.0	195.0	0.1407	830.1	1957.7	2787.8	2.284	4.181	6.465
16.0	201.4	0.1237	858.5	1933.2	2791.7	2.344	4.074	6.418
18.0	207.1	0.1103	884.5	1910.3	2794.8	2.398	3.978	6.375
20.0	212.4	0.0995	908.6	1888.7	2797.3	2.447	3.890	6.337
22.0	217.2	0.0907	930.9	1868.2	2799.1	2.492	3.809	6.301
24.0	221.8	0.0832	951.9	1848.5	2800.4	2.534	3.735	6.269
26.0	226.0	0.0768	971.7	1829.7	2801.4	2.574	3.665	6.239
28.0	230.0	0.0714	990.5	1811.5	2802.0	2.611	3.600	6.211
30.0	233.8	0.0666	1008.3	1794.0	2802.3	2.646	3.538	6.184
32.0	237.4	0.0624	1025.4	1776.9	2802.3	2.679	3.480	6.159
34.0	240.9	0.0587	1041.8	1760.3	2802.1	2.710	3.424	6.134
36.0	244.2	0.0554	1057.5	1744.2	2801.7	2.740	3.371	6.111
38.0	247.3	0.0524	1072.7	1728.4	2801.1	2.769	3.321	6.090
40.0	250.3	0.0497	1087.4	1712.9	2800.3	2.797	3.272	6.069
42.0	253.2	0.0473	1101.6	1697.8	2799.4	2.823	3.225	6.048
44.0	256.1	0.0451	1115.4	1682.9	2798.3	2.849	3.180	6.029
46.0	258.8	0.0430	1128.8	1668.3	2797.1	2.874	3.136	6.010
48.0	261.4	0.0411	1141.8	1653.9	2795.7	2.897	3.094	5.991
50.0	263.9	0.0394	1154.5	1639.7	2794.2	2.921	3.053	5.974
52.0	266.4	0.0378	1166.9	1625.7	2792.6	2.943	3.013	5.956
54.0	268.8	0.0363	1179.0	1611.8	2790.8	2.965	2.974	5.939
56.0	271.1	0.0349	1190.8	1598.2	2789.0	2.986	2.936	5.923
58.0	273.4	0.0336	1202.4	1584.6	2787.0	3.007	2.899	5.906
60.0	275.6	0.0324	1213.7	1571.3	2785.0	3.027	2.863	5.891
62.0	277.7	0.0313	1224.9	1558.0	2782.9	3.047	2.828	5.875
64.0	279.8	0.0302	1235.8	1544.8	2780.6	3.066	2.794	5.860
66.0	281.9	0.0292	1246.5	1531.8	2778.3	3.085	2.760	5.845
68.0	283.9	0.0283	1257.1	1518.8	2775.9	3.104	2.727	5.831
70.0	285.8	0.0274	1267.5	1506.0	2773.4	3.122	2.694	5.816
72.0	287.7	0.0265	1277.7	1493.2	2770.9	3.140	2.662	5.802
74.0	289.6	0.0257	1287.8	1480.5	2768.2	3.157	2.631	5.788
76.0	291.4	0.0249	1297.7	1467.8	2765.5	3.174	2.600	5.774
78.0	293.2	0.0242	1307.5	1455.3	2762.7	3.191	2.569	5.760

p	t	v_g	h_f	h_{fg}	h_g	s_f	s_{fg}	s_g
80.0	295.0	0.0235	1317.2	1442.7	2759.9	3.208	2.539	5.747
82.0	296.7	0.0228	1326.7	1430.3	2757.0	3.224	2.510	5.734
84.0	298.4	0.0222	1336.2	1417.8	2754.0	3.240	2.481	5.721
86.0	300.1	0.0216	1345.4	1405.5	2750.9	3.256	2.452	5.708
88.0	301.7	0.0210	1354.7	1393.1	2747.8	3.271	2.423	5.694
90.0	303.3	0.0205	1363.8	1380.8	2744.6	3.287	2.395	5.682
92.0	304.9	0.0200	1372.8	1368.5	2741.3	3.302	2.367	5.669
94.0	306.5	0.0195	1381.7	1356.3	2738.0	3.317	2.340	5.657
96.0	308.0	0.0190	1390.6	1344.1	2734.7	3.332	2.313	5.645
98.0	309.5	0.0185	1399.4	1331.9	2731.3	3.346	2.286	5.632
100.0	311.0	0.0180	1408.1	1319.7	2727.8	3.361	2.259	5.620
110.0	318.0	0.0160	1450.6	1258.8	2709.4	3.431	2.129	5.560
120.0	324.6	0.0143	1491.7	1197.5	2698.2	3.497	2.003	5.500
130.0	330.8	0.0128	1531.9	1135.1	2667.0	3.561	1.880	5.441
140.0	336.6	0.0115	1571.5	1070.9	2642.4	3.624	1.756	5.380
150.0	342.1	0.0103	1610.9	1004.2	2615.1	3.686	1.632	5.318
160.0	347.3	0.0093	1650.4	934.5	2584.9	3.747	1.506	5.253
170.0	352.3	0.0084	1691.6	860.0	2551.6	3.811	1.375	5.186
180.0	357.0	0.0075	1734.8	779.0	2513.8	3.877	1.236	5.113
190.0	361.4	0.0067	1778.7	691.8	2470.5	3.943	1.090	5.033
200.0	365.7	0.0059	1826.6	591.6	2418.2	4.015	0.926	4.941
210.0	369.8	0.0050	1886.3	461.2	2347.5	4.105	0.717	4.822
220.0	373.7	0.0037	2010.3	186.3	2196.6	4.293	0.288	4.581
221.2	**374.15**	**0.0032**	**2107.4**	**000.0**	**2107.4**	**4.443**	**0.000**	**4.443**

TABLE 2

Dry Saturated Steam (Temperature based)

Saturation temperature in °C	Pressure in bar	Specific volume of steam in m³/kg	Specific enthalpy in kJ/kg			Specific entropy in kJ/kg K		
			Water	Evaporation	Steam	Water	Evaporation	Steam
(t)	(p)	(v_g)	(h_f)	(h_{fg})	(h_g)	(s_f)	(s_{fg})	(s_g)
0	**0.006 11**	**206.16**	**0.0**	**2501.6**	**2501.6**	**0.000**	**9.158**	**9.158**
1	0.006 57	192.61	4.2	2499.2	2503.4	0.015	9.116	9.131
2	0.007 06	179.92	8.4	2496.8	2505.2	0.031	9.074	9.105
3	0.007 58	168.17	12.6	2494.5	2507.1	0.046	9.033	9.079
4	0.008 13	157.27	16.8	2492.1	2508.9	0.061	8.992	9.053
5	0.008 72	147.16	21.0	2489.7	2510.7	0.076	8.951	9.027
6	0.009 35	137.78	25.2	2487.4	2512.6	0.091	8.910	9.001
7	0.010 01	129.06	29.4	2485.0	2514.4	0.106	8.870	8.976
8	0.010 72	120.97	33.6	2482.6	2516.2	0.121	8.830	8.951
9	0.011 47	113.44	37.8	2480.3	2518.1	0.136	8.790	8.926
10	0.012 27	106.43	42.0	2477.9	2519.9	0.151	8.751	8.902
11	0.013 12	99.909	46.2	2475.5	2521.7	0.166	8.712	8.878
12	0.014 01	93.835	50.4	2473.2	2523.6	0.181	8.673	8.854
13	0.014 97	88.176	54.6	2470.8	2525.4	0.195	8.635	8.830
14	0.015 97	82.900	58.8	2468.5	2527.2	0.210	8.596	8.806
15	0.017 04	77.978	62.9	2466.1	2529.1	0.224	8.558	8.782
16	0.018 17	73.384	67.1	2463.8	2530.9	0.239	8.520	8.759
17	0.019 36	69.095	71.3	2461.4	2532.7	0.253	8.483	8.736
18	0.020 62	65.087	75.5	2459.0	2534.5	0.268	8.446	8.714
19	0.021 96	61.341	79.7	2456.7	2536.4	0.282	8.409	8.691
20	0.023 37	57.838	83.9	2454.3	2538.2	0.296	8.372	8.668
22	0.026 42	51.492	92.2	2449.6	2541.8	0.325	8.299	8.624
24	0.029 82	45.926	100.6	2444.9	2545.5	0.353	8.228	8.581
26	0.033 60	41.034	108.9	2440.2	2549.1	0.381	8.157	8.538
28	0.037 78	36.728	117.3	2435.4	2552.7	0.409	8.087	8.496
30	0.042 42	32.929	125.7	2430.7	2556.4	0.437	8.018	8.455
32	0.047 53	29.572	134.1	2425.9	2560.0	0.464	7.950	8.414
34	0.053 18	26.601	142.4	2421.2	2563.6	0.491	7.883	8.374
36	0.059 40	23.967	150.7	2416.4	2567.1	0.518	7.816	8.334
38	0.066 24	21.627	159.1	2411.7	2570.8	0.545	7.751	8.296
40	0.073 75	19.546	167.5	2406.9	2574.4	0.572	7.686	8.258
42	0.081 99	17.692	175.8	2402.1	2577.9	0.599	7.622	8.221
44	0.091 00	16.036	184.2	2397.3	2581.5	0.625	7.559	8.184
46	0.1009	14.557	192.6	2392.5	2585.1	0.651	7.497	8.148

t	p	v_g	h_f	h_{fg}	h_g	s_f	s_{fg}	s_g
48	0.1116	13.233	200.9	2387.7	2588.6	0.678	7.435	8.113
50	0.1234	12.046	209.3	2382.9	2592.2	0.704	7.374	8.078
52	0.1361	10.890	217.6	2378.1	2595.7	0.729	7.314	8.043
54	0.1500	10.022	226.0	2373.2	2599.2	0.755	7.254	8.009
56	0.1651	9.1587	234.4	2368.4	2602.8	0.780	7.196	7.976
58	0.1815	8.3808	242.7	2363.5	2606.2	0.806	7.137	7.943
60	0.1992	7.6785	251.1	2358.6	2609.7	0.831	7.080	7.911
62	0.2184	7.0437	259.5	2353.7	2613.2	0.856	7.023	7.879
64	0.2391	6.4690	267.8	2348.8	2616.6	0.881	6.967	7.848
66	0.2615	5.9482	276.2	2343.9	2620.1	0.906	6.911	7.817
68	0.2856	5.4756	284.6	2338.9	2623.5	0.930	6.856	7.786
70	0.3116	5.0463	293.0	2334.0	2627.0	0.955	6.802	7.757
72	0.3396	4.6557	301.4	2329.0	2630.3	0.979	6.748	7.727
74	0.3696	4.3000	309.7	2324.0	2633.7	1.003	6.695	7.698
76	0.4019	3.9757	318.2	2318.9	2637.1	1.027	6.642	7.669
78	0.4365	3.6796	326.5	2313.9	2640.4	1.051	6.590	7.641
80	0.4736	3.4091	334.9	2308.8	2643.7	1.075	6.538	7.613
82	0.5133	3.1616	343.3	2303.8	2647.1	1.099	6.487	7.586
84	0.5557	2.9350	351.7	2298.6	2650.4	1.123	6.436	7.559
86	0.6011	2.7272	360.1	2293.5	2653.6	1.146	6.386	7.532
88	0.6495	2.5365	368.5	2288.4	2656.9	1.169	6.337	7.506
90	0.7011	2.3613	376.9	2283.2	2660.1	1.193	6.287	7.480
92	0.7561	2.2002	385.4	2278.0	2663.4	1.216	6.239	7.454
94	0.8146	2.0519	393.8	2272.8	2666.6	1.239	6.190	7.429
96	0.8769	1.9153	402.2	2267.5	2669.7	1.261	6.143	7.404
98	0.9430	1.7893	410.6	2262.2	2672.8	1.285	6.095	7.380
100	**1.013**	**1.6730**	**419.1**	**2256.9**	**2676.0**	**1.307**	**6.048**	**7.355**
110	1.433	1.2099	461.3	2230.0	2691.3	1.419	5.820	7.239
120	1.985	0.891 52	503.7	2202.3	2706.0	1.528	5.601	7.129
130	2.701	0.668 14	546.3	2173.6	2719.9	1.634	5.392	7.026
140	3.614	0.508 50	589.1	2144.0	2733.1	1.739	5.189	6.928
150	4.760	0.392 45	632.1	2113.2	2745.3	1.842	4.994	6.836
160	6.181	0.306 76	675.5	2081.3	2756.7	1.943	4.805	6.748
170	7.920	0.242 55	719.2	2047.9	2767.1	2.042	4.621	6.663
180	10.027	0.193 80	763.1	2013.2	2776.3	2.139	4.443	6.582
190	12.551	0.156 32	807.5	1976.7	2784.2	2.236	4.268	6.504
200	15.549	0.127 16	852.3	1938.6	2790.9	2.331	4.097	6.428
210	19.077	0.104 24	897.7	1898.5	2796.2	2.425	3.929	6.354
220	23.198	0.086 04	943.7	1856.2	2799.9	2.518	3.764	6.282

t	p	v_g	h_f	h_{fg}	h_g	s_f	s_{fg}	s_g
230	27.976	0.071 45	990.3	1811.7	2802.0	2.610	3.601	6.211
240	33.478	0.059 65	1037.6	1764.6	2802.2	2.702	3.439	6.141
250	39.776	0.050 04	1085.8	1714.7	2800.5	2.794	3.277	6.071
260	46.943	0.042 13	1134.9	1661.5	2796.4	2.885	3.116	6.001
270	55.058	0.035 59	1185.2	1604.6	2789.8	2.976	2.954	5.930
280	64.202	0.030 13	1236.8	1543.6	2780.4	3.068	2.790	5.858
290	74.461	0.025 54	1290.0	1477.6	2767.6	3.161	2.624	5.785
300	85.927	0.021 65	1345.1	1406.0	2751.1	3.255	2.453	5.708
310	98.700	0.018 33	1402.4	1327.6	2730.0	3.351	2.277	5.628
320	112.89	0.015 48	1462.6	1241.1	2703.7	3.450	2.092	5.542
330	128.63	0.012 99	1526.5	1143.6	2670.1	3.553	1.896	5.449
340	146.05	0.010 78	1595.5	1030.7	2626.2	3.662	1.681	5.343
350	165.35	0.008 80	1671.9	895.7	2567.6	3.780	1.438	5.218
360	186.75	0.006 94	1764.2	721.3	2485.5	3.921	1.139	5.060
370	210.54	0.004 97	1890.2	452.6	2342.8	4.111	0.704	4.814
374.15	**221.20**	**0.003 17**	**2107.4**	**0.0**	**2107.4**	**4.443**	**0.000**	**4.443**

TABLE 3

Saturated Trichloro-monofluoro-methane (CCl_3F), R-11

Datum at – 40° C, h_f = 0, s_f = 0

Saturation temp. in °C	*Saturation pressure in bar*	*Specific volume in m^3 / kg*		*Specific enthalpy in kJ / kg*			*Specific entropy in kJ / kg K*	
		Liquid	*Vapour*	*Liquid*	*Vapour*	*Latent*	*Liquid*	*Vapour*
(t)	(p)	(v_f)	(v_g)	(h_f)	(h_g)	(h_{fg})	(s_f)	(s_g)
–40	**0.050 96**	**0.000 617**	**2.7625**	**0.00**	**203.48**	**203.48**	**0.0000**	**0.8730**
–38	0.059 42	0.000 619	2.5360	1.65	204.48	202.83	0.0068	0.8701
–36	0.067 87	0.000 621	2.3094	3.30	205.48	202.18	0.0137	0.8671
–34	0.076 33	0.000 623	2.0827	4.95	206.48	201.53	0.0206	0.8642
–32	0.084 78	0.000 624	1.8562	6.60	207.48	200.88	0.0274	0.8613
–30	0.093 23	0.000 626	1.6296	8.25	208.48	200.23	0.0343	0.8583
–28	0.103 45	0.000 627	1.4448	9.90	209.48	199.58	0.0410	0.8554
–26	0.115 86	0.000 629	1.3123	11.55	210.49	198.94	0.0477	0.8533
–24	0.128 28	0.000 631	1.1798	13.20	211.49	198.29	0.0544	0.8508
–22	0.142 92	0.000 633	1.0722	14.85	212.50	197.65	0.0611	0.8483
–20	0.158 69	0.000 634	0.9770	16.51	213.50	196.99	0.0678	0.8462
–18	0.174 14	0.000 636	0.8793	18.19	214.52	196.33	0.0741	0.8441
–16	0.192 80	0.000 638	0.8045	19.82	215.53	195.71	0.0808	0.8420
–14	0.212 60	0.000 639	0.7345	21.49	216.54	195.05	0.0875	0.8403
–12	0.233 90	0.000 641	0.6713	23.17	217.55	194.38	0.0938	0.8387
–10	0.258 48	0.000 643	0.6191	24.83	218.56	193.73	0.1009	0.8366
–8	0.283 06	0.000 645	0.5671	26.50	219.57	193.07	0.1063	0.8349
–6	0.302 80	0.000 647	0.5194	28.17	220.59	192.42	0.1126	0.8332
–4	0.339 67	0.000 649	0.4801	29.85	221.61	191.76	0.1189	0.8315
–2	0.369 83	0.000 651	0.4409	31.54	222.63	191.09	0.1252	0.8303
0	0.403 60	0.000 653	0.4069	33.22	223.64	190.42	0.1315	0.8286
1	0.421 90	0.000 653	0.3920	34.06	224.14	190.08	0.1344	0.8280
2	0.440 21	0.000 654	0.3770	34.90	224.65	189.75	0.1374	0.8274
3	0.458 52	0.000 655	0.3620	35.74	225.15	189.41	0.1405	0.8267
4	0.476 83	0.000 656	0.3471	36.59	225.65	189.06	0.1436	0.8261
5	0.497 20	0.000 657	0.3340	37.43	226.16	188.73	0.1465	0.8254
6	0.519 17	0.000 658	0.3225	38.28	226.66	188.38	0.1498	0.8248
7	0.541 18	0.000 659	0.3109	39.13	227.17	188.04	0.1526	0.8242
8	0.563 17	0.000 660	0.2994	39.98	227.68	187.70	0.1562	0.8236
9	0.585 17	0.000 661	0.2878	40.83	228.18	187.35	0.1586	0.8229
10	0.607 17	0.000 662	0.2763	41.68	228.69	187.01	0.1616	0.8223
11	0.633 20	0.000 663	0.2673	42.53	229.20	186.67	0.1646	0.8219

t	p	v_f	v_g	h_f	h_g	h_{fg}	s_f	s_g
12	0.659 21	0.000 664	0.2584	43.39	229.71	186.32	0.1675	0.8215
13	0.685 23	0.000 665	0.2495	44.24	230.21	185.97	0.1706	0.8210
14	0.711 24	0.000 666	0.2405	45.10	230.72	185.62	0.1738	0.8206
15	0.737 27	0.000 667	0.2316	45.95	231.22	185.27	0.1767	0.8200
16	0.765 52	0.000 669	0.2235	46.81	231.73	184.92	0.1796	0.8194
17	0.796 55	0.000 670	0.2163	47.67	232.23	184.56	0.1825	0.8170
18	0.827 59	0.000 671	0.2092	48.54	232.73	184.19	0.1855	0.8186
19	0.858 62	0.000 672	0.2021	49.40	233.24	183.84	0.1885	0.8180
20	0.889 66	0.000 673	0.1950	50.26	233.74	183.48	0.1913	0.8173
21	0.920 70	0.000 674	0.1878	51.12	234.24	183.12	0.1943	0.8169
22	0.956 25	0.000 675	0.1821	51.99	234.74	182.75	0.1972	0.8165
23	0.992 37	0.000 676	0.1764	52.85	235.24	182.39	0.2003	0.8163
24	1.024 96	0.000 677	0.1708	53.72	235.75	182.03	0.2031	0.8160
25	1.061 17	0.000 678	0.1651	54.59	236.25	181.66	0.2061	0.8157
26	1.100 74	0.000 679	0.1595	55.46	236.75	181.29	0.2089	0.8152
27	1.138 80	0.000 680	0.1542	56.32	237.25	180.93	0.2120	0.8150
28	1.179 17	0.000 681	0.1495	57.20	237.76	180.56	0.2148	0.8148
29	1.222 73	0.000 682	0.1452	58.07	238.25	180.18	0.2177	0.8144
30	1.260 70	0.000 684	0.1401	58.94	238.76	179.82	0.2206	0.8140
31	1.306 65	0.000 685	0.1362	59.82	239.25	179.43	0.2234	0.8135
32	1.348 21	0.000 686	0.1316	60.70	239.75	179.05	0.2261	0.8131
33	1.395 68	0.000 687	0.1278	61.57	240.24	178.67	0.2290	0.8128
34	1.444 14	0.000 689	0.1242	62.45	240.73	178.28	0.2320	0.8126
35	1.492 76	0.000 690	0.1205	63.32	241.23	177.91	0.2349	0.8123
36	1.541 30	0.000 691	0.1168	64.20	241.72	177.52	0.2378	0.8119
37	1.589 83	0.000 692	0.1131	65.07	242.22	177.15	0.2406	0.8116
38	1.640 00	0.000 693	0.1096	65.95	242.72	176.77	0.2433	0.8114
39	1.695 71	0.000 694	0.1066	66.84	243.20	176.36	0.2462	0.8112
40	1.751 45	0.000 696	0.1037	67.74	243.69	175.95	0.2490	0.8110
41	1.807 18	0.000 697	0.1007	68.63	244.18	175.55	0.2518	0.8108
42	1.862 92	0.000 698	0.0977	69.52	244.67	175.15	0.2546	0.8106
43	1.918 66	0.000 699	0.0947	70.41	245.16	174.75	0.2574	0.8104
44	1.979 52	0.000 700	0.0921	71.31	245.65	174.34	0.2600	0.8102
45	2.043 00	0.000 701	0.0896	72.20	246.14	173.94	0.2029	0.8100
46	2.106 41	0.000 702	0.0871	73.08	246.63	173.55	0.2658	0.8098
47	2.169 80	0.000 704	0.0847	73.98	247.12	173.14	0.2686	0.8096
48	2.233 10	0.000 705	0.0823	74.87	247.60	172.73	0.2713	0.8094
49	2.296 70	0.000 706	0.0798	75.76	248.07	172.31	0.2741	0.8092
50	2.360 00	0.000 707	0.0774	76.67	248.53	171.86	0.2768	0.8090

TABLE 4

Saturated Dichloro-difluoro-methane (CCl_2F_2), R-12

Datum at –40° C, h_f = 0, s_f = 0

Saturation temp. in °C	*Saturation pressure in bar*	*Specific volume in m^3/kg*		*Specific enthalpy in kJ / kg*			*Specific entropy in kJ / kg K*	
		Liquid	*Vapour*	*Liquid*	*Vapour*	*Latent*	*Liquid*	*Vapour*
(t)	(p)	(v_f)	(v_g)	(h_f)	(h_g)	(h_{fg})	(s_f)	(s_g)
–100	0.011 85	0.000 600	10.1951	–51.84	142.00	193.84	–0.2567	0.8628
–95	0.018 64	0.000 604	6.6231	–47.56	144.22	191.78	–0.2323	0.8442
–90	0.028 43	0.000 608	4.4206	–43.28	146.46	189.74	–0.2086	0.8273
–85	0.042 54	0.000 613	3.0531	–39.00	148.73	187.73	–0.1856	0.8122
–80	0.062 00	0.000 617	2.1519	–34.72	151.02	185.74	–0.1631	0.7985
–75	0.088 26	0.000 622	1.5462	–30.42	153.32	183.74	–0.1412	0.7861
–70	0.122 98	0.000 627	1.1314	–26.12	155.63	181.75	–0.1198	0.7665
–65	0.168 07	0.000 632	0.8421	–21.81	157.96	179.77	–0.0988	0.7648
–60	0.226 65	0.000 637	0.6401	–17.48	160.29	177.77	–0.0783	0.7558
–55	0.300 52	0.000 643	0.4930	–13.14	162.62	175.76	–0.0581	0.7475
–50	0.392 37	0.000 648	0.3845	–8.78	164.95	173.73	–0.0384	0.7401
–45	0.505 12	0.000 654	0.3035	–4.39	167.27	171.66	–0.0190	0.7334
–40	**0.641 90**	**0.000 660**	**0.2422**	**0.00**	**169.60**	**169.60**	**0.0000**	**0.7274**
–38	0.704 60	0.000 663	0.2221	1.76	170.52	168.76	0.0075	0.7251
–36	0.771 96	0.000 665	0.2040	3.53	171.44	167.91	0.0149	0.7230
–34	0.844 21	0.000 667	0.1877	5.31	172.36	167.05	0.0224	0.7209
–32	0.927 76	0.000 670	0.1729	7.08	173.28	166.20	0.0298	0.7190
–30	1.004 41	0.000 673	0.1596	8.86	174.20	165.34	0.0371	0.7171
–28	1.093 11	0.000 676	0.1475	10.64	175.11	164.47	0.0444	0.7153
–26	1.187 78	0.000 678	0.1364	12.43	176.02	163.59	0.0516	0.7135
–24	1.288 58	0.000 681	0.1265	14.22	176.93	162.71	0.0588	0.7118
–22	1.395 81	0.000 683	0.1173	16.02	177.83	161.81	0.0660	0.7102
–20	1.509 72	0.000 686	0.1090	17.82	178.73	160.91	0.0731	0.7087
–18	1.631 04	0.000 689	0.1014	19.62	179.63	160.01	0.0801	0.7073
–16	1.759 63	0.000 692	0.0944	21.43	180.53	159.10	0.0871	0.7059
–14	1.895 75	0.000 695	0.0880	23.23	181.42	159.19	0.0941	0.7045
–12	2.046 05	0.000 698	0.0821	25.05	182.31	157.26	0.1010	0.7032
–10	2.191 72	0.000 701	0.0767	26.87	183.19	156.32	0.1080	0.7019
–8	2.352 72	0.000 704	0.0717	28.70	184.06	155.36	0.1148	0.7007
–6	2.522 44	0.000 707	0.0672	30.53	184.94	154.41	0.1217	0.6996
–4	2.701 16	0.000 710	0.0630	32.37	185.80	153.43	0.1285	0.6986
–2	2.889 21	0.000 713	0.0591	34.20	186.67	152.47	0.1352	0.6975

t	p	v_f	v_g	h_f	h_g	h_{fg}	s_f	s_g
0	3.086 90	0.000 717	0.0555	36.05	187.53	151.48	0.1420	0.6965
1	3.189 74	0.000 719	0.0538	36.98	187.95	150.97	0.1453	0.6961
2	3.295 13	0.000 720	0.0521	37.90	188.38	150.48	0.1489	0.6956
3	3.403 10	0.000 721	0.0505	38.83	188.81	149.98	0.1521	0.6951
4	3.513 67	0.000 723	0.0490	39.76	189.23	149.47	0.1553	0.6947
5	3.626 90	0.000 725	0.0475	40.69	189.65	148.96	0.1586	0.6943
6	3.742 80	0.000 727	0.0461	41.62	190.07	148.45	0.1620	0.6938
7	3.861 41	0.000 729	0.0447	42.56	190.49	147.93	0.1653	0.6933
8	3.982 83	0.000 730	0.0434	43.50	190.91	147.41	0.1686	0.6929
9	4.107 02	0.000 732	0.0422	44.43	191.32	146.89	0.1719	0.6925
10	4.234 07	0.000 734	0.0410	45.37	191.74	146.37	0.1752	0.6921
11	4.364 42	0.000 736	0.0398	46.31	192.15	145.84	0.1784	0.6917
12	4.497 63	0.000 738	0.0386	47.26	192.56	145.30	0.1817	0.6913
13	4.633 86	0.000 739	0.0375	48.20	192.97	144.77	0.1850	0.6909
14	4.773 12	0.000 741	0.0365	49.15	193.38	144.23	0.1883	0.6906
15	4.915 45	0.000 743	0.0355	50.10	193.79	143.69	0.1915	0.6902
16	5.060 87	0.000 745	0.0345	51.05	194.19	143.14	0.1948	0.6898
17	5.209 42	0.000 747	0.0335	52.00	194.59	142.59	0.1981	0.6894
18	5.361 17	0.000 749	0.0326	52.95	194.99	142.04	0.2013	0.6891
19	5.516 14	0.000 751	0.0317	53.91	195.38	141.47	0.2046	0.6888
20	5.674 41	0.000 753	0.0308	54.87	195.78	140.91	0.2078	0.6884
21	5.836 35	0.000 756	0.0300	55.83	196.17	140.34	0.2110	0.6881
22	6.001 71	0.000 758	0.0292	56.79	196.56	139.77	0.2143	0.6878
23	6.170 50	0.000 759	0.0284	57.75	196.96	139.21	0.2174	0.6875
24	6.342 69	0.000 761	0.0276	58.73	197.34	138.61	0.2207	0.6872
25	6.518 40	0.000 764	0.0269	59.70	197.73	138.03	0.2239	0.6868
26	6.697 65	0.000 766	0.0262	60.67	198.11	137.44	0.2271	0.6865
27	6.880 48	0.000 768	0.0255	61.65	198.50	136.85	0.2303	0.6862
28	7.067 04	0.000 770	0.0248	62.63	198.87	136.24	0.2335	0.6859
29	7.257 38	0.000 772	0.0241	63.61	199.25	135.64	0.2368	0.6856
30	7.451 03	0.000 775	0.0235	64.59	199.62	135.03	0.2400	0.6853
31	7.649 03	0.000 777	0.0230	65.58	199.99	134.41	0.2431	0.6850
32	7.850 89	0.000 779	0.0225	66.57	200.36	133.79	0.2463	0.6847
33	8.056 62	0.000 782	0.0218	67.56	200.73	133.17	0.2495	0.6845
34	8.266 21	0.000 784	0.0212	68.56	201.09	132.53	0.2527	0.6842
35	8.480 00	0.000 786	0.0207	69.56	201.45	131.89	0.2559	0.6839
36	8.697 66	0.000 789	0.0202	70.55	201.80	131.25	0.2591	0.6836
37	8.919 04	0.000 792	0.0196	71.55	202.16	130.61	0.2623	0.6833
38	9.144 83	0.000 794	0.0191	72.56	202.51	129.95	0.2654	0.6830

t	p	v_f	v_g	h_f	h_g	h_{fg}	s_f	s_g
39	9.374 97	0.000 796	0.0186	73.57	202.86	129.29	0.2685	0.6828
40	9.608 97	0.000 799	0.0182	74.59	203.20	128.61	0.2718	0.6825
41	9.847 93	0.000 802	0.0177	75.61	203.54	127.93	0.2750	0.6822
42	10.091 31	0.000 804	0.0173	76.62	203.87	127.25	0.2782	0.6820
43	10.338 62	0.000 807	0.0168	77.65	204.21	126.56	0.2814	0.6817
44	10.590 21	0.000 810	0.0164	78.68	204.55	125.87	0.2846	0.6814
45	10.846 55	0.000 813	0.0160	79.71	204.87	125.16	0.2878	0.6811
46	11.107 58	0.000 815	0.0156	80.75	205.19	124.44	0.2909	0.6808
47	11.373 04	0.000 818	0.0153	81.79	205.51	123.72	0.2941	0.6805
48	11.642 90	0.000 821	0.0149	82.83	205.83	123.00	0.2973	0.6802
49	11.917 24	0.000 824	0.0146	83.88	206.14	122.26	0.3005	0.6800
50	12.196 55	0.000 827	0.0142	84.94	206.45	121.51	0.3037	0.6797

TABLE 5

Saturated Monochloro-difluoro-methane ($CHClF_2$) R-22

Datum at −40° C, h_f = 0, s_f = 0

Saturation temp. in °C	*Saturation pressure in bar*	*Specific volume in m^3 / kg*		*Specific enthalpy in kJ / kg*			*Specific entropy in kJ / kg K*	
		Liquid	*Vapour*	*Liquid*	*Vapour*	*Latent*	*Liquid*	*Vapour*
(t)	(p)	(v_f)	(v_g)	(h_f)	(h_g)	(h_{fg})	(s_f)	(s_g)
−100	0.020 09	0.000 643	8.3412	−63.45	203.73	267.18	−0.3144	1.2293
−95	0.031 50	0.000 647	5.4344	−58.14	206.19	264.33	−0.2843	1.2004
−90	0.047 92	0.000 652	3.6381	−52.87	208.64	261.51	−0.2550	1.1736
−85	0.077 31	0.000 656	2.5204	−47.61	211.11	258.72	−0.2269	1.1489
−80	0.103 93	0.000 661	1.7816	−42.40	213.60	256.00	−0.1989	1.1267
−75	0.147 59	0.000 666	1.2842	−37.17	216.11	253.28	−0.1721	1.1066
−70	0.205 17	0.000 672	0.9420	−31.93	218.62	250.55	−0.1461	1.0874
−65	0.279 65	0.000 677	0.7037	−26.68	221.16	247.84	−0.1206	1.0702
−60	0.374 48	0.000 683	0.5351	−21.42	223.67	245.09	−0.0959	1.0543
−55	0.496 21	0.000 689	0.4131	−16.13	226.18	242.31	−0.0712	1.0396
−50	0.647 58	0.000 696	0.3229	−10.81	228.69	239.50	−0.0473	1.0262
−45	0.832 41	0.000 702	0.2556	−5.40	231.20	236.60	−0.0234	1.0137
−40	**1.055 86**	**0.000 709**	**0.2049**	**0.00**	**233.67**	**233.67**	**0.0000**	**1.0024**
−38	1.158 62	0.000 712	0.1882	2.20	234.65	232.45	0.0096	0.9982
−36	1.269 10	0.000 715	0.1728	4.40	235.65	231.25	0.0188	0.9940
−34	1.387 31	0.000 719	0.1590	6.58	236.60	230.02	0.0280	0.9902
−32	1.513 24	0.000 721	0.1465	8.83	237.60	228.77	0.0373	0.9860
−30	1.646 90	0.000 724	0.1353	11.05	238.55	227.50	0.0464	0.9283
−28	1.789 38	0.000 727	0.1253	13.29	239.52	226.23	0.0557	0.9785
−26	1.940 69	0.000 731	0.1161	15.58	240.48	224.90	0.0645	0.9747
−24	2.102 07	0.000 734	0.1077	17.77	241.41	223.64	0.0733	0.9710
−22	2.274 48	0.000 738	0.1000	19.99	242.33	222.34	0.0821	0.9676
−20	2.457 93	0.000 741	0.0930	22.21	243.25	221.04	0.0908	0.9638
−18	2.653 10	0.000 744	0.0865	24.47	244.17	219.70	0.0996	0.9605
−16	2.859 03	0.000 748	0.0806	26.72	245.08	218.36	0.1080	0.9576
−14	3.078 76	0.000 752	0.0752	28.94	245.96	217.02	0.1164	0.9538
−12	3.311 72	0.000 756	0.0701	31.16	246.84	215.68	0.1248	0.9508
−10	3.557 93	0.000 759	0.0655	33.40	247.72	214.32	0.1336	0.9479
−8	3.813 21	0.000 763	0.0612	35.66	248.60	212.94	0.1419	0.9454
−6	4.091 72	0.000 767	0.0573	37.92	249.46	211.54	0.1503	0.9425
−4	4.379 72	0.000 771	0.0536	40.19	250.30	210.11	0.1591	0.9396
−2	4.683 17	0.000 775	0.0503	42.52	251.14	208.62	0.1675	0.9370

t	p	v_f	v_g	h_f	h_g	h_{fg}	s_f	s_g
0	5.002 07	0.000 779	0.0471	44.94	251.97	207.03	0.1763	0.9345
1	5.168 41	0.000 781	0.0457	46.16	252.37	206.21	0.1809	0.9333
2	5.339 17	0.000 783	0.0443	47.38	252.77	205.39	0.1855	0.9320
3	5.514 34	0.000 785	0.0429	48.64	253.17	204.53	0.1901	0.9303
4	5.693 52	0.000 787	0.0416	49.91	253.56	203.65	0.1943	0.9291
5	5.876 21	0.000790	0.0403	51.16	253.95	202.79	0.1989	0.9278
6	6.062 76	0.000 792	0.0391	52.41	254.33	201.92	0.2035	0.9266
7	6.253 93	0.000 794	0.0379	53.68	254.71	201.03	0.2077	0.9253
8	6.449 93	0.000 797	0.0368	54.94	255.08	200.14	0.2123	0.9241
9	6.650 34	0.000 799	0.0357	56.22	255.44	199.22	0.2169	0.9228
10	6.855 17	0.000 801	0.0346	57.52	255.81	198.29	0.2211	0.9216
11	7.066 21	0.000 803	0.0336	58.80	256.15	197.35	0.2257	0.9203
12	7.277 24	0.000 805	0.0326	60.07	256.48	196.41	0.2303	0.9190
13	7.497 93	0.000 808	0.0316	61.38	256.83	195.45	0.2345	0.9178
14	7.725 52	0.000 810	0.0307	62.72	257.17	194.45	0.2391	0.9165
15	7.955 17	0.000 813	0.0298	64.02	257.50	193.48	0.2437	0.9152
16	8.187 58	0.000 815	0.0289	65.32	257.81	192.48	0.2483	0.9140
17	8.425 52	0.000 817	0.0281	66.63	258.13	191.50	0.2529	0.9127
18	8.632 41	0.000 820	0.0273	67.95	258.44	190.49	0.2571	0.9115
19	8.917 93	0.000 822	0.0266	69.27	258.73	189.46	0.2617	0.9102
20	9.172 41	0.000 825	0.0258	70.59	259.00	188.41	0.2663	0.9089
21	9.433 10	0.000 827	0.0251	71.93	259.29	187.36	0.2709	0.9077
22	9.699 31	0.000 830	0.0244	73.31	259.58	186.27	0.2755	0.9065
23	9.971 03	0.000 833	0.0237	74.66	259.85	185.19	0.2801	0.9052
24	10.248 28	0.000 835	0.0230	76.04	260.11	184.07	0.2847	0.9039
25	10.531 03	0.000 838	0.0224	77.39	260.38	182.99	0.2889	0.9027
26	10.819 31	0.000 841	0.0218	78.79	260.64	181.85	0.2935	0.9014
27	11.113 10	0.000 844	0.0212	80.16	260.89	180.73	0.2981	0.9002
28	11.412 41	0.000 846	0.0206	81.54	261.12	179.58	0.3023	0.8989
29	11.690 34	0.000 849	0.0200	82.96	261.37	178.41	0.3069	0.8977
30	12.034 48	0.000 852	0.0194	84.38	261.60	177.22	0.3115	0.8964
31	12.351 03	0.000 855	0.0189	85.77	261.81	176.04	0.3161	0.8948
32	12.673 10	0.000 858	0.0184	87.17	262.02	174.85	0.3207	0.8935
33	13.000 70	0.000 861	0.0179	88.57	262.22	173.65	0.3249	0.8922
34	13.337 93	0.000 864	0.0174	90.00	262.40	172.40	0.3295	0.8909
35	13.682 76	0.000 867	0.0169	91.43	262.58	171.15	0.3337	0.8893
36	14.030 34	0.000 870	0.0165	92.85	262.74	169.89	0.3383	0.8880
37	14.382 07	0.000 874	0.0161	94.24	262.88	168.64	0.3429	0.8868
38	14.743 45	0.000 877	0.0156	95.63	263.00	167.37	0.3471	0.8851
39	15.111 03	0.000 881	0.0152	97.03	263.13	166.10	0.3517	0.8838

t	p	v_f	v_g	h_f	h_g	h_{fg}	s_f	s_g
40	15.489 65	0.000 884	0.0148	98.44	263.21	164.77	0.3563	0.8822
41	15.868 27	0.000 887	0.0144	99.82	263.29	163.47	0.3601	0.8809
42	16.257 93	0.000 891	0.0140	101.24	263.38	162.14	0.3651	0.8797
43	16.653 80	0.000 894	0.0137	102.68	263.48	160.80	0.3693	0.8780
44	17.055 17	0.000 898	0.0133	104.12	263.58	159.46	0.3739	0.8767
45	17.465 52	0.000 902	0.0130	105.58	263.67	158.09	0.3785	0.8755
46	17.884 14	0.000 906	0.0126	107.04	263.76	156.72	0.3827	0.8738
47	18.308 27	0.000 910	0.0123	108.51	263.87	155.36	0.3871	0.8723
48	18.737 93	0.000 914	0.0119	109.98	263.93	153.95	0.3915	0.8709
49	19.124 14	0.000 918	0.0116	111.42	263.99	152.57	0.3959	0.8694
50	19.613 80	0.000 922	0.0113	112.86	264.05	151.19	0.4003	0.8680

TABLE 6

Saturated Methyl Chloride (CH_3Cl), R-40

Datum at –40° C, h_f = 0, s_f = 0

Saturation temp. in °C	*Saturation pressure in bar*	*Specific volume in m^3 / kg*		*Specific enthalpy in kJ/kg*			*Specific entropy in kJ / kg K*	
		Liquid	*Vapour*	*Liquid*	*Vapour*	*Latent*	*Liquid*	*Vapour*
(*t*)	(*p*)	(v_f)	(v_g)	(h_f)	(h_g)	(h_{fg})	(s_f)	(s_g)
–40	**0.474 34**	**0.000 971**	**0.7950**	**0.00**	**443.48**	**443.48**	**0.0000**	**1.9026**
–38	0.527 93	0.000 974	0.7310	2.98	444.66	441.68	0.0126	1.8917
–36	0.581 52	0.000 977	0.6669	5.97	445.85	439.88	0.0255	1.8808
–34	0.637 93	0.000 981	0.6064	8.95	447.03	438.08	0.0377	1.8703
–32	0.704 34	0.000 984	0.5590	11.95	448.22	436.27	0.0502	1.8603
–30	0.770 69	0.000 988	0.5114	14.95	449.41	434.46	0.0628	1.8498
–28	0.843 45	0.000 991	0.4692	17.96	450.57	432.61	0.0749	1.8402
–26	0.855 17	0.000 994	0.4336	20.98	451.73	430.75	0.0871	1.8310
–24	1.004 83	0.000 998	0.3979	24.00	452.87	428.87	0.0992	1.8213
–22	1.096 55	0.001 001	0.3679	27.03	454.04	427.01	0.1114	1.8126
–20	1.188 96	0.001 004	0.3383	30.08	455.17	425.09	0.1235	1.8033
–18	1.315 45	0.001 009	0.3080	33.81	456.56	422.75	0.1382	1.7929
–16	1.401 65	0.001 013	0.2906	36.19	457.43	421.24	0.1474	1.7862
–14	1.519 72	0.001 016	0.2700	39.24	458.57	419.33	0.1595	1.7782
–12	1.642 07	0.001 020	0.2506	42.30	459.69	417.39	0.1712	1.7698
–10	1.773 80	0.001 023	0.2329	45.38	460.76	415.38	0.1830	1.7619
–8	1.907 60	0.001 027	0.2179	48.47	461.80	413.33	0.1947	1.7544
–6	2.061 38	0.001 031	0.2027	51.58	462.83	411.25	0.2064	1.7464
–4	2.226 90	0.001 035	0.1898	54.69	463.83	409.14	0.2177	1.7389
–2	2.389 00	0.001 039	0.1769	57.79	464.83	407.04	0.2290	1.7309
0	2.559 31	0.001 043	0.1650	60.89	465.80	404.91	0.2408	1.7238
1	2.650 55	0.001 045	0.1597	62.46	466.29	403.83	0.2466	1.7200
2	2.744 00	0.001 048	0.1546	64.02	466.77	402.75	0.2520	1.7163
3	2.840 14	0.001 050	0.1497	65.58	467.25	401.67	0.2579	1.7129
4	2.938 62	0.001 052	0.1450	67.15	467.72	400.57	0.2638	1.7096
5	3.040 00	0.001 054	0.1404	68.72	468.19	399.47	0.2696	1.7058
6	3.144 14	0.001 056	0.1360	70.29	468.67	398.38	0.2749	1.7024
7	3.250 48	0.001 058	0.1317	71.86	469.12	397.26	0.2803	1.6991

t	p	v_f	v_g	h_f	h_g	h_{fg}	s_f	s_g
8	3.359 31	0.001 060	0.1277	73.43	469.57	396.14	0.2860	1.6957
9	3.471 31	0.001 062	0.1237	75.00	470.04	395.04	0.2916	1.6924
10	3.585 52	0.001 065	0.1200	76.60	470.50	393.90	0.2973	1.6890
11	3.702 83	0.001 067	0.1164	78.16	470.94	392.78	0.3029	1.6857
12	3.823 86	0.001 069	0.1129	79.75	471.40	391.65	0.3086	1.6828
13	3.946 90	0.001 072	0.1096	81.33	471.83	390.50	0.3140	1.6794
14	4.073 80	0.001 074	0.1063	82.91	472.27	389.36	0.3194	1.6760
15	4.202 76	0.001 076	0.1032	84.50	472.71	388.21	0.3251	1.6727
16	4.336 41	0.001 079	0.1002	86.09	473.14	387.05	0.3308	1.6698
17	4.472 65	0.001 081	0.0973	87.68	473.57	385.89	0.3362	1.6664
18	4.610 48	0.001 083	0.0945	89.28	474.00	384.72	0.3416	1.6631
19	4.753 86	0.001 085	0.0918	90.89	474.43	383.54	0.3471	1.6601
20	4.897 24	0.001 087	0.0892	92.48	474.85	382.37	0.3525	1.6572
21	5.047 76	0.001 090	0.0867	94.08	475.25	381.17	0.3578	1.6539
22	5.198 28	0.001 092	0.0842	95.68	475.64	379.96	0.3631	1.6509
23	5.354 34	0.001 094	0.0819	97.29	476.04	378.75	0.3685	1.6476
24	5.512 00	0.001 096	0.0796	98.89	476.43	377.54	0.3739	1.6446
25	5.673 80	0.001 099	0.0774	100.49	476.83	376.34	0.3793	1.6417
26	5.838 90	0.001 101	0.0753	102.11	477.20	375.09	0.3848	1.6388
27	6.006 48	0.001 103	0.0733	103.72	477.57	373.85	0.3903	1.6358
28	6.179 03	0.001 105	0.0713	105.33	477.94	372.61	0.3956	1.6329
29	6.352 27	0.001 108	0.0694	106.94	478.31	371.37	0.4009	1.6300
30	6.531 03	0.001 111	0.0676	108.55	478.69	370.14	0.4061	1.6275
31	6.713 52	0.001 114	0.0658	110.17	479.03	368.86	0.4114	1.6248
32	6.900 96	0.001 116	0.0641	111.79	479.38	367.59	0.4167	1.6218
33	7.089 03	0.001 119	0.0624	113.41	479.74	366.33	0.4220	1.6191
34	7.284 14	0.001 121	0.0608	115.03	480.08	365.05	0.4274	1.6164
35	7.482 76	0.001 124	0.0592	116.65	480.41	363.76	0.4328	1.6136
36	7.684 14	0.001 126	0.0577	118.27	480.75	362.48	0.4382	1.6108
37	7.888 96	0.001 128	0.0562	119.90	481.08	361.18	0.4435	1.6082
38	8.095 86	0.001 131	0.0548	121.53	481.41	359.88	0.4487	1.6055
39	8.310 00	0.001 133	0.0534	123.16	481.74	358.58	0.4438	1.6029
40	8.524 14	0.001 136	0.0520	124.79	482.06	357.27	0.4589	1.6003
41	8.747 58	0.001 139	0.0508	126.43	482.38	355.95	0.4642	1.5976

t	p	v_f	v_g	h_f	h_g	h_{fg}	s_f	s_g
42	8.971 03	0.001 142	0.0495	128.07	482.69	354.62	0.4694	1.5950
43	9.201 72	0.001 144	0.0483	129.71	483.00	353.29	0.4746	1.5923
44	9.434 48	0.001 147	0.0472	131.34	483.31	351.96	0.4797	1.5897
45	9.674 96	0.001 150	0.0461	132.98	483.60	350.62	0.4848	1.5872
46	9.921 72	0.001 153	0.0450	134.63	483.88	349.25	0.4900	1.5848
47	10.168 27	0.001 156	0.0439	136.28	484.17	347.89	0.4952	1.5823
48	10.415 17	0.001 159	0.0429	137.93	484.46	346.53	0.5003	1.5798
49	10.664 41	0.001 162	0.0418	139.58	484.74	345.16	0.5055	1.5774
50	10.933 80	0.001 165	0.0409	141.23	485.01	343.78	0.5106	1.5750
51	11.203 17	0.001 168	0.0400	142.88	485.27	342.39	0.5158	1.5726
52	11.472 55	0.001 172	0.0391	144.55	485.53	340.98	0.5209	1.5702
53	11.741 93	0.001 175	0.0381	146.21	485.80	339.59	0.5260	1.5677
54	12.011 31	0.001 178	0.0372	147.86	486.06	338.20	0.5311	1.5653
55	12.294 48	0.001 181	0.0364	149.53	486.31	336.78	0.5362	1.5630
56	12.588 69	0.001 184	0.0356	151.19	486.55	335.36	0.5413	1.5606
57	12.882 90	0.001 187	0.0348	152.86	486.78	333.92	0.5463	1.5583
58	13.177 10	0.001 190	0.0340	154.53	487.01	332.48	0.5514	1.5560
59	13.471 31	0.001 193	0.0332	156.20	487.25	331.05	0.5564	1.5536
60	13.765 52	0.001 197	0.0324	157.86	487.48	329.62	0.5615	1.5513

TABLE 7

Saturated Trichloro-trifluoro-ethane ($C_2Cl_3F_3$), R-113

Datumn at –30°C, h_f = 0, s_f = 0

Saturation temp. in °C	*Saturation pressure in bar*	*Specific volume in m^3 / kg*		*Specific enthalpy in kJ/kg*			*Specific entropy in kJ / kg K*	
		Liquid	*Vapour*	*Liquid*	*Vapour*	*Latent*	*Liquid*	*Vapour*
(t)	(p)	(v_f)	(v_g)	(h_f)	(h_g)	(h_{fg})	(s_f)	(s_g)
–30	**0.0282**	**0.5925**	**3.7834**	**0.00**	**166.89**	**166.89**	**0.0000**	**0.6868**
–28	0.0321	0.5941	3.3693	1.75	168.11	166.36	0.0072	0.6862
–26	0.0364	0.5956	3.0047	3.50	169.33	165.83	0.0144	0.6857
–24	0.0411	0.5972	2.6839	5.26	170.56	165.30	0.0215	0.6853
–22	0.0463	0.5988	2.4016	7.03	171.79	164.76	0.0285	0.6850
–20	0.0521	0.6004	2.1529	8.80	173.02	164.22	0.0355	0.6846
–18	0.0584	0.6020	1.9336	10.57	174.25	163.68	0.0425	0.6844
–16	0.0654	0.6036	1.7398	12.35	175.49	163.14	0.0494	0.6842
–14	0.0730	0.6052	1.5685	14.13	176.73	162.60	0.0563	0.6841
–12	0.0814	0.6069	1.4167	15.92	177.96	162.04	0.0632	0.6840
–10	0.0905	0.6085	1.2820	17.71	179.20	161.49	0.0700	0.6840
–8	0.1005	0.6102	1.1622	19.51	180.45	160.94	0.0768	0.6841
–6	0.1114	0.6118	1.0555	21.31	181.69	160.38	0.0835	0.6842
–4	0.1232	0.6135	0.9602	23.12	182.94	159.82	0.0902	0.6843
–2	0.1361	0.6152	0.8751	24.93	184.19	159.26	0.0969	0.6845
0	0.1500	0.6169	0.7988	26.75	185.44	158.69	0.1035	0.6848
2	0.1651	0.6186	0.7304	28.58	186.70	158.12	0.1101	0.6851
4	0.1851	0.6204	0.6689	30.40	187.96	157.56	0.1167	0.6855
6	0.1991	0.6221	0.6135	32.24	189.22	156.98	0.1233	0.6860
8	0.2181	0.6239	0.5635	34.08	190.49	156.41	0.1298	0.6864
10	0.2386	0.6257	0.5184	35.92	191.75	155.83	0.1363	0.6870
12	0.2606	0.6274	0.4775	37.77	193.03	155.26	0.1428	0.6876
14	0.2842	0.6293	0.4405	39.62	194.30	154.67	0.1493	0.6882
16	0.3095	0.6311	0.4069	41.49	195.58	154.09	0.1557	0.6889
18	0.3366	0.6329	0.3763	43.36	196.88	153.50	0.1621	0.6896
20	0.3656	0.6348	0.3485	45.23	198.15	152.92	0.1685	0.6904
22	0.3966	0.6367	0.3231	47.11	194.44	152.33	0.1749	0.6913

t	p	v_f	v_g	h_f	h_g	h_{fg}	s_f	s_g
24	0.4296	0.6385	0.3000	49.00	200.73	151.73	0.1813	0.6922
26	0.4648	0.6405	0.2788	50.89	202.03	151.14	0.1876	0.6931
28	0.5023	0.6424	0.2594	52.79	203.33	150.54	0.1940	0.6941
30	0.5421	0.6443	0.2416	54.69	204.63	149.94	0.2003	0.6951
32	0.5844	0.6463	0.2253	56.60	205.94	149.34	0.2066	0.6962
34	0.6292	0.6483	0.2103	58.52	207.26	148.74	0.2128	0.6973
36	0.6788	0.6503	0.1965	60.44	208.57	148.13	0.2191	0.6985
38	0.7271	0.6523	0.1838	62.37	209.90	147.53	0.2253	0.6997
40	0.7803	0.6544	0.1721	64.31	211.22	146.91	0.2316	0.7010
42	0.8366	0.6564	0.1612	66.26	212.56	146.30	0.2378	0.7022
44	0.8960	0.6585	0.1512	68.21	213.90	145.69	0.2440	0.7036
46	0.9587	0.6606	0.1420	70.16	215.23	145.07	0.2502	0.7050
48	1.0247	0.6628	0.1334	72.13	216.58	144.45	0.2564	0.7064
50	1.0943	0.6649	0.1255	74.10	217.93	143.83	0.2625	0.7073
52	1.1675	0.6671	0.1181	76.05	219.29	143.24	0.2687	0.7093
54	1.2444	0.6693	0.1112	78.07	220.65	142.58	0.2748	0.7108
56	1.3253	0.6715	0.1049	80.07	222.02	141.95	0.2809	0.7124
58	1.4102	0.6738	0.0989	82.07	223.39	141.32	0.2871	0.7140
60	1.4992	0.6761	0.0934	84.09	224.77	140.68	0.2932	0.7156
62	1.5926	0.6784	0.0883	86.11	226.15	140.04	0.2992	0.7173
64	1.6904	0.6807	0.0835	88.13	227.54	139.41	0.3053	0.7190
66	1.7928	0.6830	0.0790	90.17	228.93	138.76	0.3114	0.7207
68	1.9000	0.6854	0.0748	92.22	230.34	138.12	0.3174	0.7225
70	2.0120	0.6878	0.0708	94.27	231.74	137.47	0.3234	0.7242
72	2.1291	0.6902	0.0672	96.33	233.15	136.82	0.3295	0.7260
74	2.2513	0.6972	0.0637	98.40	234.57	136.17	0.3355	0.7279
76	2.3789	0.6952	0.0605	100.48	235.99	135.51	0.3414	0.7297
78	2.5120	0.6977	0.0575	102.58	237.43	134.85	0.3474	0.7316
80	2.6508	0.7002	0.0546	104.67	238.86	134.19	0.3534	0.7335

TABLE 8

Saturated Dichloro-tetrafluoro-ethane ($C_2Cl_2F_4$), R-114

Datumn at −40°C, h_f = 0, s_f = 0

Saturation temp. in °C	*Saturation pressure in bar*	*Specific volume in m^3 / kg*		*Specific enthalpy in kJ/kg*			*Specific entropy in kJ / kgK*	
(*t*)	(*p*)	*Liquid* (v_f)	*Vapour* (v_g)	*Liquid* (h_f)	*Vapour* (h_g)	*Latent* (h_{fg})	*Liquid* (s_f)	*Vapour* (s_g)
−40	0.1305	0.6061	0.8351	0.00	146.77	146.77	0.0000	0.6299
−38	0.1471	0.6080	0.7557	1.51	147.95	146.44	0.0065	0.6296
−36	0.1649	0.6100	0.6849	3.04	149.14	146.10	0.0129	0.6293
−34	0.1841	0.6120	0.6216	4.58	150.33	145.75	0.0192	0.6291
−32	0.2049	0.6141	0.5650	6.14	151.52	145.38	0.0256	0.6288
−30	0.2273	0.6161	0.5142	7.71	152.72	145.01	0.0319	0.6287
−28	0.2516	0.6182	0.4686	9.29	153.92	144.63	0.0382	0.6285
−26	0.2779	0.6203	0.4276	10.88	155.12	144.24	0.0445	0.6284
−24	0.3064	0.6224	0.3907	12.49	156.33	143.84	0.0507	0.6284
−22	0.3372	0.6245	0.3575	14.12	157.54	143.42	0.0569	0.6283
−20	0.3705	0.6266	0.3275	15.76	158.75	142.99	0.0632	0.6284
−18	0.4065	0.6288	0.3004	17.42	159.97	142.55	0.0694	0.6284
−16	0.4453	0.6310	0.2758	19.09	161.19	142.10	0.0756	0.6285
−14	0.4870	0.6332	0.2536	20.79	162.42	141.63	0.0818	0.6287
−12	0.5320	0.6354	0.2335	22.50	163.65	141.15	0.0881	0.6289
−10	0.5803	0.6377	0.2152	24.22	164.88	140.66	0.0943	0.6291
−8	0.6322	0.6400	0.1986	25.96	166.11	140.15	0.1005	0.6294
−6	0.6877	0.6423	0.1835	27.72	167.35	139.63	0.1068	0.6297
−4	0.7472	0.6446	0.1697	29.50	168.59	139.09	0.1130	0.6301
−2	0.8107	0.6470	0.1571	31.29	169.83	138.54	0.1193	0.6305
0	0.8766	0.6494	0.1456	33.11	171.08	137.97	0.1255	0.6309
2	0.9500	0.6518	0.1351	34.94	172.33	137.39	0.1318	0.6314
4	1.0279	0.6542	0.1255	36.79	173.59	136.80	0.1381	0.6320
6	1.1098	0.6567	0.1168	38.66	174.84	136.18	0.1444	0.6325
8	1.1967	0.6592	0.1087	40.54	176.10	135.56	0.1507	0.6332

t	p	v_f	v_g	h_f	h_g	h_{fg}	s_f	s_g
10	1.2889	0.6617	0.1013	42.45	177.37	134.92	0.1571	0.6338
12	1.3866	0.6642	0.0945	44.38	178.64	134.26	0.1634	0.6345
14	1.4900	0.6668	0.0882	46.32	179.91	133.59	0.1698	0.6352
16	1.5993	0.6695	0.0825	48.28	181.18	132.90	0.1761	0.6360
18	1.7147	0.6721	0.0772	50.26	182.46	132.20	0.1825	0.6368
20	1.8365	0.6748	0.0722	52.25	183.73	131.48	0.1888	0.6376
22	1.9648	0.6775	0.0678	54.27	185.02	130.75	0.1952	0.6384
24	2.1000	0.6803	0.0636	56.30	186.30	130.00	0.2015	0.6393
26	2.2421	0.6831	0.0597	58.35	187.59	129.24	0.2079	0.6401
28	2.3915	0.6859	0.0561	60.43	188.89	128.46	0.2143	0.6410
30	2.5483	0.6888	0.0528	62.51	190.18	127.67	0.2206	0.6419
32	2.7129	0.6918	0.0498	64.62	191.48	126.86	0.2269	0.6429
34	2.8854	0.6947	0.0469	66.74	192.78	126.04	0.2332	0.6438
36	3.0661	0.6977	0.0443	68.88	194.09	125.21	0.2395	0.6447
38	3.2552	0.7008	0.0418	71.04	195.40	124.36	0.2458	0.6456
40	3.4530	0.7039	0.0395	73.21	196.71	123.50	0.2520	0.6466
42	3.6697	0.7071	0.0374	75.40	198.03	122.63	0.2582	0.6475
44	3.8755	0.7103	0.0354	77.60	199.35	121.75	0.2643	0.6484
46	4.1009	0.7136	0.0335	79.82	200.67	120.85	0.2704	0.6492
48	4.3358	0.7169	0.0318	82.05	201.99	119.94	0.2764	0.6501
50	4.5808	0.7202	0.0302	84.30	203.32	119.02	0.2824	0.6509
52	4.8359	0.7237	0.0287	86.57	204.66	118.09	0.2883	0.6517
54	5.1015	0.7272	0.0272	88.84	205.99	117.15	0.2942	0.6524
56	5.3778	0.7307	0.0259	91.13	207.33	116.20	0.2999	0.6531
58	5.6651	0.7343	0.0247	93.44	208.68	115.24	0.3056	0.6538
60	5.9637	0.7380	0.0235	95.75	210.02	114.27	0.3112	0.6544

TABLE 9

Saturated Dichloro-trifluoro-ethane (CF_3 CH Cl_2), R-123

Datumn at 0°C; h_f = 200 kJ/kg ; s_f = 1.000 kJ/kg K

Saturation temp. in °C	*Saturation pressure in bar*	*Specific volume in m^3/kg*		*Specific enthalpy in kJ/kg*			*Specific entropy in kJ / kg K*	
		Liquid	*Vapour*	*Liquid*	*Vapour*	*Latent*	*Liquid*	*Vapour*
(t)	(p)	(v_f)	(v_g)	(h_f)	(h_g)	(h_{fg})	(s_f)	(s_g)
–40	0.0378	0.000 620	3.3365	167.23	356.30	189.07	0.8704	1.6814
–30	0.0698	0.000 628	1.8798	175.47	362.04	186.57	0.9050	1.6723
–20	0.1221	0.000 637	1.1145	183.38	367.87	184.49	0.9370	1.6657
–10	0.2041	0.000 646	0.6904	191.48	373.77	182.29	0.9683	1.6610
0	**0.3273**	**0.000 656**	**0.4445**	**200.00**	**379.75**	**179.75**	**1.0000**	**1.6581**
2	0.3518	0.000 658	0.4088	201.76	380.95	179.19	1.0064	1.6577
4	0.3912	0.000 660	0.3764	203.54	382.16	178.62	1.0129	1.6573
6	0.4267	0.000 662	0.3471	205.34	383.37	178.03	1.0193	1.6571
8	0.4648	0.000 664	0.3204	207.16	384.58	177.42	1.0258	1.6568
10	0.5057	0.000 666	0.2962	209.00	385.80	176.80	1.0323	1.6567
12	0.5494	0.000 669	0.2742	210.86	387.00	176.14	1.0389	1.6566
14	0.5961	0.000 671	0.2540	212.74	388.22	175.48	1.0454	1.6565
16	0.6459	0.000 673	0.2357	214.64	389.43	174.79	1.0520	1.6565
18	0.6990	0.000 675	0.2190	216.55	390.65	174.10	1.0586	1.6565
20	0.7555	0.000 677	0.2035	218.50	391.87	173.37	1.0652	1.6566
22	0.8157	0.000 680	0.1894	220.44	393.10	172.66	1.0718	1.6568
24	0.8796	0.000 682	0.1765	222.40	394.31	171.91	1.0784	1.6569
26	0.9473	0.000 684	0.1646	224.40	395.54	171.14	1.0850	1.6572
28	1.0192	0.000 687	0.1537	226.38	396.76	170.38	1.0917	1.6574
30	1.0952	0.000 689	0.1437	228.40	397.98	169.58	1.0983	1.6577
32	1.1757	0.000 692	0.1345	230.42	399.21	168.79	1.1050	1.6581
34	1.2607	0.000 694	0.1260	232.46	400.43	167.97	1.1116	1.6585
36	1.3504	0.000 697	0.1180	234.52	401.65	167.13	1.1183	1.6589
38	1.4451	0.000 700	0.1107	236.60	402.88	166.28	1.1249	1.6594
40	1.5448	0.000 702	0.1039	238.67	404.10	165.43	1.1316	1.6598
42	1.6498	0.000 704	0.0977	240.76	405.32	164.56	1.1382	1.6604
44	1.7603	0.000 707	0.0918	242.86	406.54	163.68	1.1448	1.6609
46	1.8764	0.000 710	0.0865	245.00	408.00	163.00	1.1514	1.6615
48	1.9983	0.000 713	0.0814	247.10	409.00	161.90	1.1580	1.6621
50	2.1261	0.000 715	0.0768	249.23	410.20	160.97	1.1646	1.6627
52	2.2602	0.000 718	0.0724	251.37	411.40	160.03	1.1712	1.6634
54	2.4007	0.000 721	0.0684	253.50	412.60	159.10	1.1778	1.6641

t	p	v_f	v_g	h_f	h_g	h_{fg}	s_f	s_g
56	2.5478	0.000 724	0.0646	255.70	413.82	158.12	1.1843	1.6648
58	2.7016	0.000 727	0.0611	257.85	415.02	157.17	1.1909	1.6655
60	2.8624	0.000 730	0.0578	260.03	416.22	156.19	1.1974	1.6662
70	3.7782	0.000 745	0.0442	271.02	422.15	151.13	1.2297	1.6701
80	4.8995	0.000 762	0.0342	282.16	427.94	145.78	1.2614	1.6742
90	6.2537	0.000 781	0.0268	293.43	433.54	140.11	1.2926	1.6784
100	7.8696	0.000 802	0.0212	304.83	438.88	134.05	1.3232	1.6825
110	9.7760	0.000 825	0.0170	316.37	443.92	127.55	1.3533	1.6862
120	12.010	0.000 852	0.0136	328.10	448.56	120.46	1.3830	1.6895
130	14.600	0.000 883	0.0110	340.03	452.72	112.69	1.4125	1.6920
140	17.587	0.000 919	0.0088	352.32	456.26	103.94	1.4419	1.6935
150	21.010	0.000 965	0.0070	365.12	458.94	93.82	1.4718	1.6935
160	24.921	0.001 026	0.0055	378.70	460.32	81.62	1.5026	1.6910

TABLE 10

Saturated Tetrafluoro-ethane (CF_3 CH_2 F), R-134(a)

Datumn at 0°C, h_f = 200 kJ/kg ; s_f = 1.000 kJ/kg K

Saturation temp. in °C	Saturation pressure in bar	Specific volume in m^3/kg		Specific enthalpy in kJ/kg			Specific entropy in kJ / kg K	
		Liquid	Vapour	Liquid	Vapour	Latent	Liquid	Vapour
(t)	(p)	(v_f)	(v_g)	(h_f)	(h_g)	(h_{fg})	(s_f)	(s_g)
–60	0.1594	0.000 680	1.0770	123.96	361.51	237.55	0.6871	1.0816
–50	0.2948	0.000 693	0.6056	136.21	367.83	231.62	0.7432	1.7812
–40	0.5122	0.000 707	0.3610	148.57	374.16	225.59	0.7973	1.7650
–30	0.8436	0.000 721	0.2260	161.10	380.45	219.35	0.8498	1.7520
–20	1.3268	0.000 737	0.1474	173.82	386.66	212.84	0.9009	1.7417
–10	2.0052	0.000 754	0.0996	186.78	392.75	205.97	0.9509	1.7337
0	**2.9270**	**0.000 773**	**0.0693**	**200.00**	**398.68**	**198.68**	**1.000**	**1.7274**
2	3.1450	0.000 777	0.0647	202.68	399.84	197.16	1.0097	1.7263
4	3.3755	0.000 781	0.0604	205.37	401.00	195.63	1.0194	1.7252
6	3.6186	0.000 785	0.0565	208.08	402.14	194.06	1.0291	1.7242
8	3.8750	0.000 789	0.0528	210.80	403.27	192.47	1.0387	1.7233
10	4.1450	0.000 793	0.0495	213.53	404.40	190.87	1.0483	1.7224
12	4.4290	0.000 798	0.0464	216.27	405.50	189.23	1.0580	1.7215
14	4.7276	0.000 802	0.0435	219.03	406.60	187.57	1.0674	1.7207
16	5.0413	0.000 807	0.0408	221.80	407.70	185.90	1.0770	1.7200
18	5.3706	0.000 812	0.0383	224.60	408.78	184.18	1.0865	1.7190
20	5.7160	0.000 816	0.0360	227.40	409.84	182.44	1.0960	1.7183
22	6.0777	0.000 821	0.0339	230.21	410.89	180.68	1.1055	1.7176
24	6.4566	0.000 826	0.0319	233.05	411.93	178.88	1.1150	1.7170
26	6.8531	0.000 831	0.0300	235.90	412.95	177.05	1.1244	1.7162
28	7.2676	0.000 837	0.0283	238.77	413.95	175.18	1.1338	1.7155
30	7.7008	0.000 842	0.0267	241.65	414.94	173.29	1.1432	1.7150
32	8.1530	0.000 848	0.0252	244.55	415.90	171.35	1.1527	1.7142
34	8.6250	0.000 854	0.0237	247.47	416.85	169.38	1.1621	1.7135
36	9.1172	0.000 860	0.0224	250.41	417.78	167.37	1.1715	1.7129
38	9.6301	0.000 866	0.0212	253.37	418.70	165.33	1.1809	1.7122
40	10.165	0.000 872	0.0200	256.35	419.58	163.23	1.1903	1.7115
42	10.721	0.000 879	0.0189	259.35	420.44	161.09	1.1997	1.7108
44	11.300	0.000 886	0.0178	262.38	421.28	158.90	1.2091	1.7101
46	11.901	0.000 893	0.0169	265.42	422.10	156.68	1.2185	1.7094

t	p	v_f	v_g	h_f	h_g	h_{fg}	s_f	s_g
48	12.527	0.000 900	0.0160	268.50	422.88	154.38	1.2279	1.7068
50	13.177	0.000 907	0.0151	271.60	423.63	152.03	1.2373	1.7078
52	13.852	0.000 915	0.0143	274.71	424.35	149.64	1.2468	1.7070
54	14.553	0.000 923	0.0135	277.86	425.03	147.17	1.2562	1.7061
56	15.290	0.000 932	0.0128	281.04	425.68	144.64	1.2657	1.7051
58	16.033	0.000 941	0.0121	284.25	426.30	142.05	1.2752	1.7041
60	16.815	0.000 950	0.0114	287.50	426.86	139.36	1.2847	1.7031
70	21.165	0.001 00	0.0087	304.30	428.90	124.60	1.3332	1.6963
80	26.331	0.001 08	0.0065	322.41	429.02	106.61	1.3837	1.6855
90	32.445	0.001 20	0.0046	343.01	425.48	82.47	1.4392	1.6663
100	39.721	0.001 54	0.0026	374.02	407.08	33.06	1.5207	1.6093

TABLE 11

Saturated Ammonia (NH_3), R-717

Datum at –40° C, $h_f = 0$, $s_f = 0$

Saturation temp. in °C	Saturation pressure in bar	Specific volume in m^3/kg		Specific enthalpy in kJ/kg			Specific entropy in kJ / kg K	
		Liquid	Vapour	Liquid	Vapour	Latent	Liquid	Vapour
(t)	(p)	(v_f)	(v_g)	(h_f)	(h_g)	(h_{fg})	(s_f)	(s_g)
–50	0.408 96	0.001 426	2.6281	–44.43	1373.27	1417.70	–0.1943	6.1603
–48	0.459 72	0.001 431	2.3565	–35.44	1376.80	1412.24	–0.1551	6.1192
–46	0.516 00	0.001 436	2.1177	–26.60	1380.20	1406.80	–0.1157	6.0789
–44	0.577 10	0.001 441	1.9062	–17.81	1383.31	1401.12	–0.0769	6.0394
–42	0.644 55	0.001 446	1.7196	–8.97	1386.67	1395.64	–0.0384	6.0008
–40	**0.717 93**	**0.001 451**	**1.5537**	**0.00**	**1390.02**	**1390.02**	**0.0000**	**5.9631**
–38	0.793 84	0.001 456	1.4077	8.97	1393.13	1384.16	0.0381	5.9262
–36	0.886 07	0.001 462	1.2775	17.81	1396.35	1378.54	0.0758	5.8900
–34	0.980 96	0.001 467	1.1614	26.84	1399.51	1372.67	0.1131	5.8545
–32	1.081 65	0.001 472	1.0574	35.68	1402.48	1366.80	0.1506	5.8198
–30	1.195 86	0.001 477	0.9644	44.66	1405.60	1360.94	0.1876	5.7856
–28	1.317 24	0.001 483	0.8820	53.68	1408.53	1354.85	0.2242	5.7521
–26	1.447 90	0.001 488	0.8069	62.61	1411.45	1348.84	0.2607	5.7195
–24	1.588 41	0.001 494	0.7397	71.73	1414.39	1342.66	0.2970	5.6872
–22	1.739 86	0.001 500	0.6793	80.76	1417.28	1336.52	0.3330	5.6556
–20	1.902 76	0.001 505	0.6244	89.78	1420.02	1330.24	0.3684	5.6244
–18	2.078 07	0.001 511	0.5750	98.76	1422.72	1323.96	0.4043	5.5939
–16	2.265 51	0.001 517	0.5303	107.83	1425.28	1317.45	0.4397	5.5639
–14	2.466 34	0.001 523	0.4896	116.95	1427.88	1310.93	0.4747	5.5356
–12	2.680 69	0.001 529	0.4526	126.16	1430.54	1304.38	0.5096	5.5055
–10	2.908 96	0.001 536	0.4189	135.37	1433.05	1297.68	0.5443	5.4770
–8	3.153 65	0.001 541	0.3884	144.35	1435.33	1290.98	0.5789	5.4487
–6	3.413 80	0.001 548	0.3604	153.56	1437.93	1284.37	0.6139	5.4210
–4	3.690 62	0.001 554	0.3348	162.77	1440.02	1277.25	0.6473	5.3940
–2	3.984 27	0.001 561	0.3113	171.98	1442.17	1270.19	0.6812	5.3670
0	4.295 86	0.001 567	0.2898	181.20	1444.45	1263.25	0.7151	5.3405
1	4.458 48	0.001 571	0.2798	185.80	1445.49	1259.69	0.7321	5.3277
2	4.626 62	0.001 574	0.2702	190.40	1446.54	1256.14	0.7487	5.3145

t	p	v_f	v_g	h_f	h_g	h_{fg}	s_f	s_g
3	4.799 31	0.001 578	0.2610	195.17	1447.59	1252.42	0.7653	5.3017
4	4.976 82	0.001 582	0.2521	199.85	1448.63	1248.78	0.7818	5.2888
5	5.158 62	0.001 585	0.2436	204.46	1449.56	1245.10	0.7989	5.2765
6	5.347 45	0.001 589	0.2354	209.06	1450.49	1241.43	0.8154	5.2638
7	5.540 07	0.001 592	0.2276	213.73	1451.54	1237.81	0.8320	5.2513
8	5.738 20	0.001 595	0.2201	218.50	1452.54	1234.04	0.8487	5.2389
9	5.941 86	0.001 598	0.2128	223.11	1453.39	1230.28	0.8652	5.2266
10	6.151 03	0.001 603	0.2060	227.72	1454.22	1226.50	0.8814	5.2141
11	6.366 41	0.001 607	0.1992	232.53	1455.30	1222.77	0.8979	5.2020
12	6.587 31	0.001 610	0.1928	237.15	1456.11	1218.96	0.9142	5.1900
13	6.814 20	0.001 614	0.1866	241.92	1456.96	1215.04	0.9307	5.1780
14	7.047 17	0.001 617	0.1807	246.60	1457.80	1211.20	0.9470	5.1659
15	7.282 76	0.001 621	0.1751	251.44	1458.63	1207.19	0.9634	5.1542
16	7.531 04	0.001 624	0.1696	256.14	1459.47	1203.33	0.9794	5.1421
17	7.781 38	0.001 628	0.1643	259.88	1460.24	1200.36	0.9956	5.1302
18	8.038 62	0.001 633	0.1592	265.54	1460.92	1195.38	1.0118	5.1186
19	8.305 51	0.001 637	0.1543	270.35	1461.75	1191.40	1.0280	5.1073
20	8.572 41	0.001 641	0.1496	275.16	1462.60	1187.44	1.0442	5.0956
21	8.851 72	0.001 645	0.1450	279.77	1463.21	1183.44	1.0604	5.0843
22	9.136 55	0.001 648	0.1407	284.56	1463.84	1179.28	1.0763	5.0729
23	9.426 90	0.001 652	0.1365	289.37	1464.63	1175.26	1.0924	5.0616
24	9.726 90	0.001 656	0.1324	294.19	1465.33	1171.14	1.1083	5.0503
25	10.027 60	0.001 661	0.1284	298.90	1465.84	1166.94	1.1242	5.0391
26	10.340 69	0.001 665	0.1246	303.82	1466.59	1162.77	1.1402	5.0279
27	10.661 37	0.001 669	0.1210	308.63	1467.22	1158.59	1.1563	5.0170
28	10.991 72	0.001 673	0.1174	313.45	1467.85	1154.40	1.1721	5.0061
29	11.327 58	0.001 678	0.1140	318.26	1468.45	1150.19	1.1879	4.9951
30	11.668 96	0.001 682	0.1107	323.08	1468.87	1145.79	1.2037	4.9842
31	12.016 55	0.001 686	0.1075	327.89	1469.50	1141.61	1.2195	4.9733
32	12.375 17	0.001 691	0.1045	332.71	1469.94	1137.23	1.2350	4.9624
33	12.744 82	0.001 695	0.1015	337.52	1470.36	1132.84	1.2508	4.9517
34	13.121 37	0.001 700	0.0987	342.48	1470.92	1128.44	1.2664	4.9409
35	13.503 45	0.001 704	0.0960	347.50	1471.43	1123.93	1.2821	4.9302

t	p	v_f	v_g	h_f	h_g	h_{fg}	s_f	s_g
36	13.893 79	0.0017 09	0.0932	352.29	1471.70	1119.41	1.2978	4.9196
37	14.295 17	0.0017 13	0.0907	357.25	1472.19	1114.94	1.3135	4.9091
38	14.706 20	0.0017 18	0.0882	362.12	1472.45	1110.33	1.3290	4.8985
39	15.122 76	0.0017 23	0.0857	367.11	1472.87	1105.76	1.3445	4.8885
40	15.544 83	0.0017 27	0.0834	371.93	1473.30	1101.37	1.3600	4.8774
41	15.985 51	0.0017 32	0.0811	376.95	1473.50	1096.55	1.3754	4.8669
42	16.431 72	0.0017 38	0.0789	381.79	1473.70	1091.91	1.3908	4.8563
43	16.883 44	0.0017 42	0.0768	386.76	1473.91	1087.15	1.4065	4.8461
44	17.344 82	0.0017 47	0.0747	391.79	1474.12	1082.33	1.4221	4.8357
45	17.817 24	0.0017 52	0.0727	396.81	1474.33	1077.52	1.4374	4.8251
46	18.300 70	0.0017 57	0.0708	401.84	1474.54	1072.70	1.4528	4.8147
47	18.795 18	0.0017 63	0.0689	406.86	1474.68	1067.82	1.4683	4.8045
48	19.299 31	0.0017 68	0.0670	411.89	1474.76	1062.87	1.4835	4.7938
49	19.809 65	0.0017 73	0.0652	416.91	1474.84	1057.93	1.4990	4.7834
50	20.331 03	0.0017 79	0.0636	421.94	1474.92	1052.98	1.5148	4.7732

TABLE 12

Saturated Carbon-dioxide (CO_2), R-744

Datum at –40° C, $h_f = 0$, $s_f = 0$

Saturation temp. in °C	Saturation pressure in bar	Specific volume in m^3 / kg		Specific enthalpy in kJ/kg			Specific entropy in kJ / kg K	
		Liquid	Vapour	Liquid	Vapour	Latent	Liquid	Vapour
(t)	(p)	(v_f)	(v_g)	(h_f)	(h_g)	(h_{fg})	(s_f)	(s_g)
–40	**10.055 17**	**0.000 898**	**0.038 21**	**0.00**	**320.52**	**320.52**	**0.0000**	**1.3754**
–38	10.849 65	0.000 904	0.035 77	3.77	320.86	317.09	0.0161	1.3654
–36	11.644 13	0.000 911	0.033 32	7.54	321.19	313.65	0.0322	1.3553
–34	12.466 76	0.000 918	0.031 00	11.32	321.51	310.19	0.0483	1.3453
–32	13.387 86	0.000 925	0.029 06	15.17	321.76	306.59	0.0641	1.3356
–30	14.308 96	0.000 932	0.027 12	19.02	322.01	302.99	0.0800	1.3259
–28	15.288 55	0.000 939	0.025 35	22.95	322.22	299.27	0.0956	1.3163
–26	16.341 24	0.000 947	0.023 79	26.97	322.39	295.42	0.1109	1.3067
–24	17.393 93	0.000 955	0.022 22	30.99	322.56	291.57	0.1264	1.2971
–22	18.542 62	0.000 963	0.020 84	35.06	322.72	287.66	0.1419	1.2879
–20	19.739 31	0.000 971	0.019 57	39.17	322.89	283.72	0.1574	1.2787
–18	20.936 00	0.000 980	0.018 30	43.27	323.06	279.79	0.1733	1.2692
–16	22.227 58	0.000 990	0.017 16	47.45	322.89	275.44	0.1888	1.2605
–14	23.583 45	0.001 000	0.016 11	51.73	322.61	270.88	0.2060	1.2523
–12	25.011 31	0.001 010	0.015 13	56.32	322.58	266.26	0.2260	1.2464
–10	26.540 69	0.001 021	0.014 26	60.85	322.24	261.39	0.2432	1.2372
–8	28.070 07	0.001 032	0.013 38	65.37	321.91	256.54	0.2600	1.2280
–6	29.660 69	0.001 043	0.012 56	70.06	321.54	251.48	0.2771	1.2192
–4	31.373 79	0.001 056	0.011 82	75.08	321.12	246.04	0.2954	1.2101
–2	33.086 89	0.001 069	0.011 10	80.10	320.71	240.61	0.3135	1.2010
0	34.910 34	0.001 082	0.010 41	85.27	320.01	234.74	0.3312	1.1909
1	35.866 20	0.001 090	0.010 10	87.91	319.55	231.64	0.3400	1.1854
2	36.822 07	0.001 098	0.009 79	90.54	319.09	228.55	0.3488	1.1799
3	37.778 00	0.001 106	0.009 47	93.18	318.63	225.45	0.3575	1.1744
4	38.733 80	0.001 113	0.009 16	95.82	318.11	222.35	0.3662	1.1690
5	39.750 34	0.001 122	0.008 87	98.55	317.57	219.02	0.3754	1.1633
6	40.815 45	0.001 131	0.008 60	101.36	316.86	215.50	0.3849	1.1575
7	41.880 55	0.001 140	0.008 34	104.16	316.14	211.98	0.3944	1.1516

t	p	v_f	v_g	h_f	h_g	h_{fg}	s_f	s_g
8	42.945 65	0.001 149	0.008 07	106.97	315.43	208.46	0.4039	1.1458
9	44.010 76	0.001 158	0.007 80	109.77	314.72	204.95	0.4134	1.1400
10	45.075 86	0.001 167	0.007 53	112.58	314.01	201.43	0.4229	1.1342
11	46.255 17	0.001 178	0.007 29	115.55	312.79	197.24	0.4329	1.1274
12	47.434 48	0.001 190	0.007 05	118.52	311.58	193.06	0.4430	1.1205
13	48.613 79	0.001 202	0.006 82	121.49	310.36	188.87	0.4534	1.1137
14	49.793 10	0.001 213	0.006 58	124.47	309.15	184.68	0.4635	1.1068
15	50.972 41	0.001 225	0.006 34	127.44	307.94	180.50	0.4735	1.1000
16	52.205 80	0.001 238	0.006 11	130.68	306.41	175.79	0.4843	1.0921
17	53.506 76	0.001 254	0.005 90	134.05	304.48	170.43	0.4954	1.0830
18	54.807 72	0.001 270	0.005 69	137.48	302.55	165.07	0.5065	1.0738
19	56.108 70	0.001 286	0.005 47	140.92	300.63	159.71	0.5176	1.0647
20	57.409 65	0.001 302	0.005 26	144.35	298.70	154.35	0.5286	1.0556
21	58.710 62	0.001 318	0.005 05	147.78	296.78	149.00	0.5397	1.0465
22	60.127 45	0.001 345	0.004 82	151.96	293.29	141.33	0.5539	1.0332
23	61.558 76	0.001 374	0.004 60	156.23	289.60	133.37	0.5681	1.0194
24	62.990 07	0.001 404	0.004 38	160.50	285.92	125.42	0.5832	1.0056
25	64.421 38	0.001 433	0.004 16	164.77	282.23	117.46	0.5978	0.9918
26	65.852 70	0.001 462	0.003 94	169.04	278.55	109.51	0.6124	0.9781
27	67.319 30	0.001 501	0.003 71	174.08	274.16	100.08	0.6289	0.9617
28	68.856 55	0.001 560	0.003 47	180.64	268.37	87.73	0.6490	0.9402
29	70.393 80	0.001 619	0.003 23	187.19	262.58	75.39	0.6691	0.9187
30	71.931 03	0.001 678	0.002 99	193.75	256.79	63.04	0.6892	0.8972
31	73.531 03	0.002 160	0.002 16	225.62	225.62	0.00	0.7913	0.7913

TABLE 13

Saturated Sulphur dioxide (SO_2), R-764

Datum at –40° C, h_f = 0, s_f = 0

Saturation temp. in °C	Saturation pressure in bar	Specific volume in m^3 / kg		Specific enthalpy in kJ/kg			Specific entropy in kJ / kg K	
		Liquid	Vapour	Liquid	Vapour	Latent	Liquid	Vapour
(t)	(p)	(v_f)	(v_g)	(h_f)	(h_g)	(h_{fg})	(s_f)	(s_g)
–40	**0.216 28**	**0.000 652**	**1.4012**	**0.00**	**415.44**	**415.44**	**0.0000**	**1.7820**
–38	0.245 94	0.000 654	1.2694	2.45	416.52	414.07	0.0101	1.7715
–36	0.275 61	0.000 656	1.1375	4.90	417.61	412.71	0.0203	1.7610
–34	0.307 31	0.000 659	1.0143	7.38	418.68	411.30	0.0305	1.7505
–32	0.345 78	0.000 661	0.9211	9.93	419.64	409.71	0.0409	1.7404
–30	0.384 32	0..000 663	0.8280	12.49	420.62	408.13	0.0514	1.7303
–28	0.427 57	0.000 665	0.7462	15.09	421.56	406.47	0.0619	1.7203
–26	0.476 73	0.000 667	0.6794	17.75	422.45	404.70	0.0726	1.7106
–24	0.525 90	0.000 669	0.6125	20.42	423.33	402.91	0.0833	1.7008
–22	0.583 45	0.000 671	0.5575	23.13	424.17	401.04	0.0941	1.6912
–20	0.645 17	0.000 673	0.5090	25.88	424.93	399.05	0.1049	1.6818
–18	0.706 93	0.000 676	0.4606	28.63	425.73	397.10	0.1158	1.6723
–16	0.776 83	0.000 678	0.4199	31.44	426.46	395.02	0.1267	1.6630
–14	0.853 24	0.000 680	0.3841	34.24	427.13	392.89	0.1376	1.6538
–12	0.935 17	0.000 682	0.3518	37.06	427.74	390.68	0.1485	1.6447
–10	1.023 45	0.000 685	0.3227	39.89	428.31	388.42	0.1594	1.6356
–8	1.118 20	0.000 687	0.2966	42.76	428.86	386.10	0.1703	1.6267
–6	1.219 72	0.000 689	0.2729	45.63	429.36	383.73	0.1812	1.6177
–4	1.327 58	0.000 692	0.2509	40.50	429.80	381.30	0.1919	1.6088
–2	1.442 89	0.000 695	0.2319	51.39	430.21	378.82	0.2028	1.6000
0	1.566 20	0.000 697	0.2148	54.26	430.54	376.28	0.2135	1.5912
1	1.630 76	0.000 699	0.2061	55.73	430.71	374.28	0.2189	1.5868
2	1.697 82	0.000 701	0.1983	57.19	430.86	373.67	0.2242	1.5824
3	1.767 83	0.000 702	0.1910	58.65	430.99	372.34	0.2295	1.5781
4	1.837 85	0.000 703	0.1837	60.11	431.11	371.00	0.2349	1.5737
5	1.911 03	0.000 704	0.1768	61.56	431.22	369.66	0.2402	1.5694
6	1.986 76	0.000 705	0.1704	63.02	431.31	368.29	0.2455	1.5651
7	2.062 48	0.000 707	0.1640	64.48	431.40	366.92	0.2509	1.5608
8	2.142 76	0.000 708	0.1581	65.94	431.48	365.54	0.2561	1.5564
9	2.224 96	0.000 710	0.1524	67.40	431.54	364.14	0.2614	1.5522
10	2.306 89	0.000 711	0.1467	68.87	431.61	362.74	0.2667	1.5479
11	2.396 27	0.000 712	0.1417	70.34	431.65	361.31	0.2719	1.5436

t	p	v_f	v_g	h_f	h_g	h_{fg}	s_f	s_g
12	2.485 65	0.000 714	0.1366	71.80	431.68	359.88	0.2771	1.5393
13	2.576 55	0.000 715	0.1317	73.27	431.70	358.43	0.2823	1.5350
14	2.672 89	0.000 716	0.1273	74.72	431.69	356.97	0.2875	1.5308
15	2.679 24	0.000 718	0.1228	76.18	431.68	355.50	0.2927	1.5266
16	2.869 10	0.000 719	0.1186	77.64	431.66	354.02	0.2978	1.5224
17	2.973 38	0.000 720	0.1146	79.09	431.62	352.53	0.3030	1.5182
18	3.077 65	0.000 722	0.1106	80.55	431.58	351.03	0.3082	1.5140
19	3.186 73	0.000 723	0.1069	82.01	431.52	349.51	0.3132	1.5097
20	3.298 20	0.000 725	0.1033	83.46	431.46	348.00	0.3183	1.5055
21	3.409 65	0.000 726	0.0997	84.92	431.39	346.47	0.3233	1.5013
22	3.529 10	0.000 728	0.0965	86.38	431.29	344.91	0.3283	1.4971
23	3.649 51	0.000 729	0.0933	87.83	431.19	343.36	0.3334	1.4929
24	3.770 92	0.000 731	0.0902	89.29	431.08	341.79	0.3384	1.4887
25	3.900 27	0.000 732	0.0873	90.74	430.94	340.20	0.3434	1.4846
26	4.029 65	0.000 734	0.0844	92.19	430.80	338.61	0.3484	1.4805
27	4.162 07	0.000 735	0.0817	93.64	430.65	337.01	0.3532	1.4764
28	4.301 24	0.000 737	0.0791	95.08	430.48	335.40	0.3580	1.4721
29	4.440 27	0.000 738	0.0765	96.53	430.31	333.78	0.3629	1.4677
30	4.582 76	0.000 740	0.0740	97.97	430.12	332.15	0.3677	1.4635
31	4.731 72	0.000 741	0.0718	99.41	429.91	330.50	0.3725	1.4593
32	4.880 69	0.000 743	0.0695	100.86	429.70	328.84	0.3773	1.4551
33	5.036 41	0.000 745	0.0673	102.29	429.47	327.18	0.3821	1.4509
34	5.194 07	0.000 746	0.0653	103.71	429.22	325.51	0.3867	1.4467
35	5.351 72	0.000 748	0.0632	105.13	428.98	323.85	0.3914	1.4425
36	5.523 53	0.000 749	0.0612	106.56	428.70	322.14	0.3961	1.4383
37	5.695 34	0.000 751	0.0593	107.98	428.43	320.45	0.4007	1.4341
38	5.871 00	0.000 753	0.0575	109.40	428.14	318.74	0.4053	1.4299
39	6.060 19	0.000 754	0.0558	110.81	427.81	317.00	0.4098	1.4256
40	6.249 38	0.000 756	0.0542	112.21	427.48	315.27	0.4143	1.4214
41	6.438 56	0.000 758	0.0526	113.62	427.15	313.53	0.4188	1.4172
42	6.627 75	0.000 759	0.0510	115.03	426.82	311.79	0.4233	1.4129
43	6.816 93	0.000 761	0.0493	116.43	426.46	310.06	0.4278	1.4087
44	7.055 20	0.000 763	0.0478	117.83	426.11	308.28	0.4322	1.4044
45	7.317 93	0.000 765	0.0465	119.22	425.71	306.49	0.4365	1.4002
46	7.580 80	0.000 767	0.0451	120.61	425.32	304.71	0.4409	1.3959
47	7.843 60	0.000 769	0.0438	122.00	424.92	302.92	0.4452	1.3916
48	8.106 40	0.000 771	0.0424	123.39	424.52	301.13	0.4495	1.3874
49	8.365 98	0.000 773	0.0411	124.78	424.11	299.33	0.4538	1.3831
50	8.599 86	0.000 775	0.0401	126.13	423.60	297.47	0.4578	1.3787

TABLE 14

Psychrometric properties of air at standard pressure of 760 mm of Hg or 1.01325 bar

Temp.	*Properties of water and steam*					*Properties of dry air at 760 mm of Hg*		*Properties of mixture of dry air and saturaled steam i.e. (1+ W) kg at total pressure of 760 mm of Hg*		
	Saturation pressure of water and steam		*Enthalpy of saturated water*	*Enthalpy of saturated steam*	*Specific volume of saturated steam*	*Specific volume*	*Enthalpy*	*Volume of mixture (1+W)kg per kg of dry air*	*Enthalpy of mixture (1+W) kg per kg of dry air*	*Specific humidity per kg of dry air*
°C	*mm of Hg*	*bar*	*kJ/kg*	*kJ/kg*	*m³/kg*	*m³/kg*	*kJ/kg*	*m³/kg*	*m³/kg*	*kg/kg of dry air*
(1)	(2)	(3)	(4)	(5)	(6)	(7)	(8)	(9)	(10)	(11)
0	4.580	0.006 11	0.0	2501.6	206.16	0.7734	0.000	0.7781	9.473	0.003 789
1	4.924	0.006 57	4.2	2503.4	192.61	0.7762	1.006	0.7813	11.204	0.004 076
2	5.291	0.007 06	8.4	2505.2	179.92	0.7791	2.012	0.7845	12.982	0.004 381
3	5.681	0.007 58	12.6	2507.1	168.17	0.7819	3.018	0.7878	14.811	0.004 707
4	6.097	0.008 13	16.8	2508.9	157.27	0.7848	4.024	0.7911	16.696	0.005 054
5	6.539	0.008 72	21.0	2510.7	147.16	0.7876	5.030	0.7944	18.639	0.005 424
6	7.010	0.009 35	25.2	2512.6	137.78	0.7904	6.036	0.7978	20.644	0.005 818
7	7.510	0.010 01	29.4	2514.4	129.06	0.7933	7.041	0.8012	22.713	0.006 237
8	8.041	0.010 72	33.6	2516.2	120.97	0.7961	8.047	0.8046	24.852	0.006 683
9	8.605	0.011 47	37.8	2518.1	113.44	0.7990	9.053	0.8081	27.064	0.007 157
10	9.204	0.012 27	42.0	2519.9	106.43	0.8018	10.059	0.8116	29.352	0.007 661
11	9.839	0.013 12	46.2	2521.7	99.909	0.8046	11.065	0.8152	31.724	0.008 197
12	10.512	0.014 01	50.4	2523.6	93.835	0.8075	12.071	0.8188	34.179	0.008 766
13	11.225	0.014 97	54.6	2525.4	88.176	0.8103	13.077	0.8225	36.726	0.009 370
14	11.982	0.015 97	58.8	2527.2	82.900	0.8132	14.084	0.8262	39.370	0.010 012
15	12.781	0.017 04	62.9	2529.1	77.978	0.8160	15.090	0.8300	42.113	0.010 692
16	13.628	0.018 17	67.1	2530.9	73.384	0.8188	16.096	0.8338	44.963	0.011 413
17	14.524	0.019 36	71.3	2532.7	69.095	0.8217	17.102	0.8377	47.926	0.012 178
18	15.470	0.020 62	75.5	2534.5	65.087	0.8245	18.108	0.8414	51.008	0.012 989
19	16.470	0.021 96	79.7	2536.4	61.341	0.8274	19.114	0.8457	54.216	0.013 848
20	17.53	0.023 37	83.9	2538.2	57.838	0.8302	20.121	0.8498	57.555	0.014 758
21	18.64	0.024 86	88.0	2540.0	54.561	0.8330	21.127	0.8540	61.035	0.015 721
22	19.82	0.026 42	92.2	2541.8	51.492	0.8359	22.133	0.8583	64.660	0.016 741

(1)	(2)	(3)	(4)	(5)	(6)	(7)	(8)	(9)	(10)	(11)
23	21.06	0.028 08	96.4	2543.6	48.619	0.8387	23.140	0.8627	68.440	0.017 821
24	22.37	0.029 82	100.6	2545.5	45.926	0.8416	24.146	0.8671	72.385	0.018 963
25	23.75	0.031 66	104.8	2547.3	43.402	0.8444	25.153	0.8717	76.500	0.020 170
26	25.50	0.033 60	108.9	2549.1	41.034	0.8472	26.159	0.8764	80.798	0.021 448
27	26.73	0.035 64	113.1	2550.9	38.813	0.8501	27.165	0.8811	85.285	0.022 798
28	28.34	0.037 78	117.3	2552.7	36.728	0.8529	28.172	0.8860	89.976	0.024 226
29	30.03	0.040 04	121.5	2554.5	34.769	0.8558	29.179	0.8910	94.878	0.025 735
30	31.81	0.042 42	125.7	2556.4	32.929	0.8586	30.185	0.8962	100.006	0.027 329
31	33.69	0.044 98	129.9	2558.2	31.199	0.8614	31.192	0.9015	105.369	0.029 014
32	35.65	0.047 53	134.1	2560.0	29.572	0.8643	32.198	0.9069	110.979	0.030 793
33	37.72	0.050 35	138.3	2561.8	28.042	0.8671	33.205	0.9125	116.857	0.032 674
34	39.89	0.053 18	142.4	2563.6	26.601	0.8700	34.212	0.9183	123.011	0.034 660
35	42.17	0.056 29	146.6	2565.4	25.245	0.8728	35.219	0.9242	129.455	0.036 756
36	44.55	0.059 40	150.7	2567.1	23.967	0.8756	36.226	0.9303	136.209	0.038 971
37	47.06	0.062 82	154.9	2569.0	22.763	0.8785	37.233	0.9366	143.290	0.041 309
38	49.69	0.066 24	159.1	2570.8	21.627	0.8813	38.239	0.9431	150.713	0.043 778
39	52.44	0.070 00	163.3	2572.6	20.557	0.8842	39.246	0.9498	158.504	0.046 386
40	55.31	0.073 75	167.5	2574.4	19.546	0.8870	40.253	0.9568	166.683	0.049 141
41	58.33	0.077 86	171.7	2576.2	18.592	0.8898	41.261	0.9640	175.265	0.052 049
42	61.49	0.081 98	175.8	2577.9	17.692	0.8927	42.268	0.9714	184.275	0.055 119
43	64.80	0.086 50	180.0	2579.7	16.841	0.8955	43.275	0.9792	193.749	0.058 365
44	68.25	0.091 00	184.2	2581.5	16.036	0.8983	44.282	0.9872	203.699	0.061 791

TABLE 15

Important Properties of Commonly Used Refrigerants

Refrigerant name	Refrigerant number	Chemical formula (°C)	Boiling temperature (°C)	Freezing temperature (°C)	Critical temperature (bar)	Critical pressure at −15°C (p_E) (bar)	Evaporator pressure at 30°C (p_c) (bar)	Condenser pressure	Compression ratio (p_c/p_E)
Trichloro- monofluoro-methane	R-11	CCl_3F	+23.77	−111	198	43.8	0.2021	1.2607	6.24
Dichloro-difluoro-methane	R-12	CCl_2F_2	−29	−157.5	112	41.2	1.8262	7.4510	4.08
Dichloro-monofluoro-methane	R-21	$CHCl_2F$	+9	−135	178.5	51.65	0.3618	2.1540	5.95
Monochloro-difluoro-methane	R-22	$CHClF_2$	−41	−160	96	49.38	2.9670	12.034	4.05
Methylene chloride	R-30	CH_2Cl_2	+39.8	−96.6	216	44.14	0.0807	0.7310	9.06
Methyl chloride	R-40	CH_3Cl	−23.7	−97.5	143	66.83	1.4586	6.531	4.47
Trichloro-trifluoro-ethane	R-113	$C_2Cl_3F_3$	+ 47.6	−35	214	34.14	0.0676	0.5421	8.02
Dichloro-tetrafluoro-ethane	R-114	$C_2Cl_2F_4$	+ 3.6	−94	145.8	32.61	0.4650	2.5161	5.41
Dichloro-trifluoro-ethane	R-123	CF_3CCl_2	+27.85	—	183.8	36.74	0.0163	0.1095	4.69
Tetrafluoro-ethane	R-134a	CF_3CH_2F	−26.15	−101	101.06	40.56	1.6397	7.7008	6.72
Ammonia	R-717	NH_3	−33.3	−77.8	133	113.86	2.3634	11.67	4.94
Carbon dioxide	R-744	CO_2	−73.6	−56.7	31	73.8	22.90	71.93	3.14
Sulphur dioxide	R-764	SO_2	−10	−75.8	157	78.7	0.8145	4.583	5.63

TABLE 16

Recommended Inside Design Conditions for Summer Comfort Cooling

Outside DBT(°C)	*Occupancy over 40 minutes*				*Occupancy below 40 minutes*			
	DBT (°C)	*WBT* (°C)	*RH* (%)	*ET* (°C)	*DBT* (°C)	*WBT* (°C)	*RH* (%)	*ET* (°C)
	23.9	18.3	60	21.7	24.4	18.9	61	22.2
26.5	25	17.2	47	21.7	25.6	17.8	47	22.2
	26.1	16.1	35	21.7	26.7	16.7	36	22.2
	24.4	18.9	61	22.2	25	19.4	61	22.8
29.5	25.6	17.8	47	22.2	26.1	18.3	48	22.8
	26.7	16.7	36	22.2	27.2	17.2	36	22.8
	25	19.4	61	22.8	25.6	20.6	64	23.3
32.5	26.1	18.3	48	22.8	26.7	19.4	52	23.3
	27.2	17.2	36	22.8	27.8	18.3	40	23.3
	25.6	20.6	64	23.3	26.1	21.1	65	23.9
35.5	26.7	19.4	52	23.3	27.2	20	52	23.9
	27.8	18.3	40	23.3	28.3	18.9	41	23.9
	26.1	21.1	65	23.9	27.2	21.7	63	24.4
37.5	27.2	20	52	23.9	28.3	20.6	50	24.4
	28.3	18.9	41	23.9	29.5	19.4	38	24.4
	26.6	21.7	65	24.2	27.2	22.2	65	24.7
40.5	27.8	20.6	52	24.2	28.3	21.1	54	24.7
	28.9	19.4	42	24.2	29.5	20	41	24.7

TABLE 17

Outside Summer Design Conditions for Important Cities in India

City	DBT (°C)	WBT (°C)	RH (%)	ET (°C)
Agra	41.5	22	17	29.7
Ahmedabad	43.2	26.4	37	32.5
Ahmednagar	42.2	31.1	45	29.5
Alibagh	31.6	25.8	63	28.4
Aligarh	42.2	28.3	35	29.7
Allahabad	41.7	25	26.5	30.8
Ambala	43.3	26.7	29	29.5
Amritsar	40.1	28.1	—	29.9
Bangalore	32.9	24.7	52	28.1
Baroda	40.3	29.1	45	31.7
Bhopal	40.2	23.2	24	29.7
Chandigarh	40.1	23.9	27	29.9
Chennai	38.5	28.6	47	25.6
Cochin	35	27.8	59	28.1
Coimbatore	34.7	27.5	58	30
Cuttack	40.6	30.6	48	32
Darjeeling	17.2	14.5	77	16.8
Dehradun	40.6	26.7	35	27.8
Delhi	40.4	23.9	26	29.9
Guwahati	30.9	25.4	62	27.6
Hyderabad	39.5	26.6	37	30.8
Indore	39.4	25.1	32	29.7
Jaipur	40.8	21.7	18	28.6
Jamnagar	37.8	27.2	44	30.6
Jamshedpur	39.4	28.4	45	31
Jodhpur	43.3	26.7	29	30.6
Kanpur	41.2	21.5	17	28.6
Kolkata	35.3	27.9	60	30.3
Lucknow	42.8	28.3	34	30.3
Ludhiana	40.1	27.9	41	29.9
Madurai	38.3	25.6	38	30.6
Mahabaleshwar	28.8	19.2	40	24.2
Mangalore	35.6	27.8	60	29
Mumbai	32.8	26.7	64	29

City	DBT (°C)	WBT (°C)	RH (%)	ET (°C)
Mysore	33.3	25.6	52	29.2
Nagpur	42.6	24.6	27	30.6
Patna	37.9	26.7	42	30.5
Pune	37.1	25.6	40	30.5
Puri	32	27.8	75	29.5
Rajpur	43.3	28.3	34	30.5
Rajkot	40.5	28.9	43	31.2
Roorkee	39	26.7	40	29.5
Shillong	29.4	21.1	48	26.6
Shrinagar	25	18.4	55	22.2
Simla	22.9	12.5	28	19.2
Surat	36.3	21.6	26	30.5
Trivandrum	30.7	29.4	90	29.7
Tiruchirappalli	38.6	25.6	32	30.2
Varansi	40.8	26.1	31	31.9
Visakhapatnam	33.3	26.9	39	29.3

TABLE 18

Recommended Storage conditions of perishable products

Product	*Range of storage temperature in °C*		*Relative humidity in percentage*		*Freezing point in °C*	*Composition in % water*	*Maximum storage period*
	Short-term storage	*Long-term storage*	*Gravity air circulation*	*Forced air circulation*			
(1)	*(2)*	*(3)*	*(4)*	*(5)*	*(6)*	*(7)*	*(8)*
Fruits							
Apples	1.7 to 4.4	–1.1 to 0	85	88	–1.94	85	8 months
Bananas	12.8 to 13.3	12.8 to 13.3	80	80	–3.3 to –1.1	75	10 days
Berries	1.7 to 4.4	–0.6 to 0.6	80	85	–3.3 to –1.7	84	7-20 days
Grapes	1.7 to 4.4	–1.1 to 0	80	85	–2.2	77	1-6 months
Grapefruit	4.4 to 7.2	0 to 1.1	85	90	–1.94	77	3 months
Lemons	12.8 to 15.6	10 to 12.8	80	85	–2.2	89	3 months
Oranges	4.4 to 7.2	0 to 1.1	80	85	–2.2	86	2 months
Peaches	1.7 to 4.4	–0.6 to 0.6	80	85	–1.4	88	1 month
Pears	1.7 to 4.4	–1.1 to 0	85	90	–1.94	84	1-7 months
Pineapples	4.4 to 7.2	3.3 to 4.4	75	85	–2.2	88	1 month
Vegetables							
Artichokes	4.4 to 7.2	–0.6 to 0.6	90	90	–2.5	92	2-5 months
Avocados	4.4 to 7.2	0.6 to 1.7	85	85	–0.6	94	10 days
Beans (green)	4.4 to 7.2	0 to 1.1	85	90	–1.1	68.5	1 month
Beans (dried)	10 to 15.6	0 to 1.7	60	60	—	12.5	12 months

CHARTS